Signals and Communication Technology

This series is devoted to fundamentals and applications of modern methods of signal processing and cutting-edge communication technologies. The main topics are information and signal theory, acoustical signal processing, image processing and multimedia systems, mobile and wireless communications, and computer and communication networks. Volumes in the series address researchers in academia and industrial R&D departments. The series is application-oriented. The level of presentation of each individual volume, however, depends on the subject and can range from practical to scientific.

Indexing: All books in "Signals and Communication Technology" are indexed by Scopus and zbMATH

For general information about this book series, comments or suggestions, please contact Mary James at mary.james@springer.com or Ramesh Nath Premnath at ramesh.premnath@springer.com.

Lei Meng
Editor

International Conference on Cloud Computing and Computer Network

CCCN 2023

Springer

Editor
Lei Meng
School of Software
Shangdong University
Jinan, China

ISSN 1860-4862 ISSN 1860-4870 (electronic)
Signals and Communication Technology
ISBN 978-3-031-47099-8 ISBN 978-3-031-47100-1 (eBook)
https://doi.org/10.1007/978-3-031-47100-1

This Springer imprint is published by the registered company Springer Nature Switzerland AG
The registered company address is: Gewerbestrasse 11, 6330 Cham, Switzerland

Paper in this product is recyclable.

Preface

Lei Meng

It's my great pleasure to introduce this volume of proceedings to you, 2023 International Conference on Cloud Computing and Computer Network (CCCN 2023). CCCN 2023 is a premier, annual forum for researchers and scholars from multiple disciplines to come together to share knowledge, discuss ideas, exchange information, and learn about cutting-edge research in diverse fields.

This year, CCCN 2023 was held online successfully during April 21–23, 2023. Though the authors and speakers from all over the world couldn't communicate face to face, with virtual platform the passion for involvement wasn't affected.

We received submissions from around 15 countries, such as China, the United States, Singapore, Austria, India, Ecuador, Philippines, Pakistan, Iran, Bangladesh, Sri Lanka, etc. After 2–3 rounds of peer-reviewing process, accepted papers of the proceedings of CCCN 2023 were divided into three parts: Digital Image Detection and Application, Machine Learning and Intelligent Applications, and Computer Models and Artificial Intelligence Algorithms.

On behalf of the Organizing Committee, we would like to thank the authors for the high-quality submission. We are grateful to all reviewers for the active involvement. We also would like to express our appreciation to the members of the CCCN 2023 conference committees. Without their contribution, the conference could not achieve a complete success. Hope to see you next year in CCCN 2024.

Conference Committees

Conference Co-Chairs
Lei Meng, Shangdong University, China
Marcus Randall, Bond University Faculty of Business, Australia

Program Chairs
Liying Zheng, Harbin Engineering University, China
Tigang Jiang, University of Electronic Science and Technology of China, China
Zaixing He, Zhejiang University, China

Program Co-chairs
Angel-Antonio San-Blas, Universidad Miguel Hernández, Spain
Xinyue Zhao, Zhejiang University, China

Publicity Chair
Lei Chen, Shandong University, China

Local Chair
Teo Tee Hui, Singapore University of Technology and Design, Singapore

Technical Committees
Yiyang Chang, Purdue University, USA
Casey How, Singapore University of Social Sciences, Singapore
Liu Ziwen, Singapore University of Social Sciences, Singapore
Aaron Tan, Technical University of Munich (TUM) Asia, Singapore
Masoud Barati, Cardiff University, UK
Qian He, University of Electronic Science and Technology of China, China
Augusto Neto, Federal University of Rio Grande do Norte, Brazil
Zhiyu Jiang, Northwestern Polytechnical University, China
Carlos Delgado, University of Alcalá, Spain
Janvier Kamanzi, Cape Peninsula University of Technology, South Africa

Chen Wang, Huazhong University of Science and Technology, China
Liu Fang, Singapore University of Social Sciences, Singapore
Psannis Konstantinos, University of Macedonia, Greece
Yilun Shang, Northumbria University, UK
June Tay, Singapore University of Social Sciences, Singapore
Jingzheng Ren, Hong Kong Polytechnic University, China
Patrizio Dazzi, Information Science and Technologies Institute (ISTI), Italy
Paul Wu Horng Jyh, Singapore University of Social Sciences, Singapore
YEW KEE WONG, Hong Kong Chu Hai College, Hong Kong, China
Zhu Yongqing, Singapore University of Social Sciences, Singapore
You Xie, Technical University of Munich, Germany

Contents

Part I
Digital Image Detection and Application

Application of Convolutional Neural Networks for the Detection of Diseases in the CCN-51 Cocoa Fruit by Means of a Mobile Application

Mauro Morales, Jerson Morocho, Ximena López, and Patricio Navas

1 Introduction

Despite the enormous effort made by the world to reduce plant loss and food security, several references [1, 2] confirm that more than 20% of crop losses in the global scenario are due to plant diseases. This problem has worsened in the last decade due to the impact of pollution and climate change. With the recent development of various agricultural technologies, farmers opt for plant disease databases or consult local pathologists via telephones, instead of the classical procedure of sending plants to the diagnostic laboratory to propose the appropriate treatment. In addition, there are many attempts to use ICT (Information technology) tools to improve the efficiency of agricultural development, taking advantage of the widespread use of mobile phones.

Regarding plant disease detection, there are many articles introducing this application using one of the standard CNN (Convolutional Neural Networks) design architectures [3], such as SqueezeNet [4], ResNeXt (Aggregated Residual Transformations for Deep Neural Networks) [5], ResNet (Deep Residual Learning for Image Recognition) [6], NiN (Network In Network) [7], GoogLeNet [8], VGGNet [9], ZFNet [10], AlexNet [11], and so on. Numerous techniques and applications have been developed to reduce crop loss due to diseases.

Cocoa is one of the most important agricultural products in world markets [12]. For this reason, the focus of this project is to detect potential threats to cocoa

M. Morales (✉) · X. López · P. Navas
University of the Armed Forces ESPE, Latacunga, Ecuador
e-mail: mjmorales4@espe.edu.ec; xrlopez@espe.edu.ec; mpnavas@espe.edu.ec

J. Morocho
Quijano and Ordoñez, Hermanas Páez, Latacunga, Cotopaxi, Ecuador
e-mail: jamorocho3@espe.edu.ec

L. Meng (ed.), *International Conference on Cloud Computing and Computer Network*, Signals and Communication Technology,
https://doi.org/10.1007/978-3-031-47100-1_1

by taking a photograph that will be processed and issued with an analysis of the fruit, based on the similarity in detecting quality in other fruits. This approach has been practical since the advent of Deep Learning, which is powerful in image classification [13, 14]. It is also the preferred method for other computer vision tasks, unlike the traditional method which is based on feature extraction algorithm such as SIFT [15], SURF [16], PCA, and LDA. This fact was first demonstrated by Girshick et al. [17] for object detection, which became popular for other computer vision tasks.

2 Background

2.1 Importance of Cocoa in Ecuador

Latin America is known as the birthplace of cocoa and recent archeological research suggests that the place of origin of cocoa is Ecuador. Ceramic remains of cocoa were found in the Amazon rainforest dating back to 3300 BC, which means that cocoa beans have been cultivated in Ecuador for more than 5000 years [18].

In 1790, with the abolition of the law prohibiting the export of cocoa, Guayaquil became the world's leading cocoa port, maintaining a monopoly that lasted almost 150 years [18]. Today, Ecuador is the leading cocoa producer in Latin America and the fifth largest in the world, according to reports from the FAO Food and Agriculture Organization of the United Nations, although the lack of technology and low resistance to pests and diseases is a limiting factor [19].

Cocoa CCN-51 A cocoa variety originating in Ecuador obtained in the 1960s by producer Homero Castro Zurita, in the canton of Naranjal, Guayas province. Among the benefits of planting this cocoa variety are its adaptability to the country's different climatic zones, high productivity with good crop management, and resistance to diseases and pests [20]. CCN-51 cocoa has organoleptic characteristics demanded by the international market, being one of the second most productive and internationally recognized varieties [20].

2.2 Cocoa Pests

Cocoa Fly Caused by the *Monalonion dissimulatum* bug, exclusive of cocoa, the insects feed on the shoots when young, and when they reach adulthood, they feed on the pods, causing pustules or circular wounds in the apical half of the fruit [21].

Bull's Horn Caused by the sucking insect *Hoplophorion pertusa*, which in its adult stage feeds on the sap of the shoots and young branches, sucking the juices from the plant with its stylet. Excessive shade in the cocoa plantation predisposes to a greater attack by the pest [21].

2.3 Cocoa Diseases

Moniliasis Caused by the fungus *Moniliophthera roreri*, it affects the fruits, having variable symptoms according to the age of the fruit, as the infection progresses, a spot with white cottony tissue appears, this tissue turns gray due to the appearance of spores or seeds, ending with the mummification and deformation of the fruit [21].

Witches' Broom Disease caused by the fungus *Crinipellis pernicosa*, it causes an abnormal sprouting at the level of both terminal and auxiliary buds, presenting a concentration of branches from a single point known as broom, in the affected floral cushions the flowers remain attached to this for a longer time than normal, developing unfertilized ovules, if the fruits are attacked, it produces malformations similar to those caused by moniliasis [21].

2.4 Technology in Agriculture

Technology goes hand in hand with farming, as farmers have always sought to make the hard work of farming easier. From the moldboard plough to tractors connected via satellite to your mobile phone, technology in agriculture means advances and improvements in the efficiency of your farm, defining technology in agriculture as an advance in the model of working and improving the efficiency and exploitation of the farm, technology should help to optimize profitability and therefore the farmer's economy [22], an example of which is the digitalization of production processes, increasing yields and saving costs; on the other hand, we have the automation of fruit and seed selection processes, applying monitoring and photogrammetry techniques in the process.

3 System Architecture

The selected architecture is based on the C4 Model where the specific model for the development of the API (Application Programming Interfaces) has been considered in its first iteration, this component will implement an onion architecture that will allow it to be tolerant to change. The respective components for the web and mobile clients are external to the implementation and depend heavily on the API, so the design of its architecture will be clearer in the next iteration (as shown in Fig. 1).

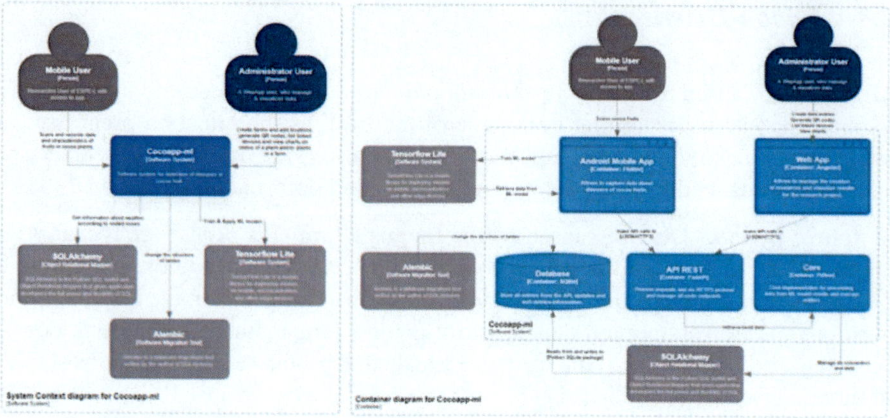

Fig. 1 First iteration of the cocoa fruit disease detection system represented by the C4 Model and Structure of the application

4 System Development

Based on the C4 model, we sought to represent the structure and different modules that the application will have, dividing the system into two main modules: the Mobile User and the administration module. In the case of the first module (Mobile User), it will be an Android application, using the Dark programming language in Flutter, which will communicate with an API in charge of sending the information emitted by it for comparison with a dataset trained by means of TensorFlow Lite.

The second module (Administrator User) will be a web application that will also communicate with the database through the REST API, facilitating this connection thanks to SQLAlchemy and SQLite using Alembic for the Data versioning.

The development of the application is divided into four modules (see Fig. 2):

Development of the Web Application Corresponding to the administration application, starting with the layout phase and subsequent development in the Angular framework for the front-end and Python in conjunction with Fast-Api for the consumption of the API-Rest, using Alembic and making the connection to SQLite.

Development of the Mobile Application Developed in Flutter, incorporating the use of QR codes to link the benefactor farm with the data collection, in addition to the creation of the CNN for image analysis.

Dataset Generation and Training This consists of the preparation of the data, which includes the selection and processing of the images for the use of the model.

Testing This stage includes the training of the neural network with the previously prepared data, the evaluation of the accuracy of the model, and the adjustment of the hyperparameters to improve its accuracy, which will later be exported for the use of the app.

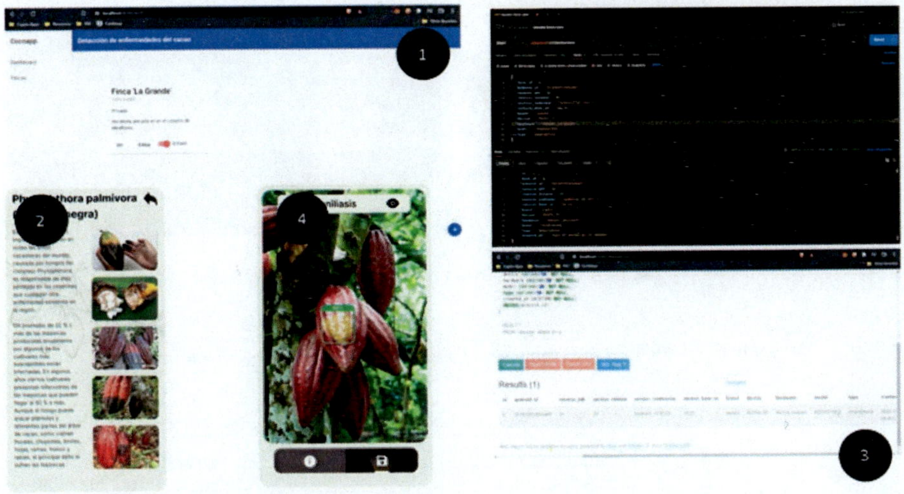

Fig. 2 Stages of development of the CACAO disease detection application

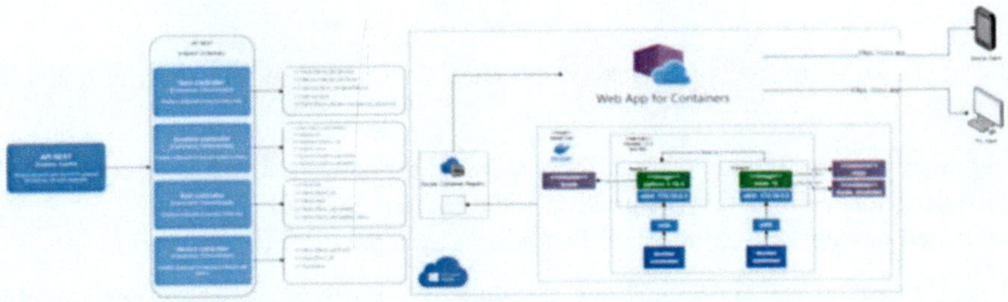

Fig. 3 Available endpoints and deployment of the application

4.1 Deploy

The application consists of a main API, in charge of receiving and providing the necessary information to manage the cocoa fruit visualization and analysis cycle, Fig. 3 shows a diagram of the available endpoints. The deployment of the project is intended to be deployed in the Azure cloud, using the "Docker Container Registry" and "Web App for containers" services through which the project can be made available on the web (Fig. 3).

5 Result and Discussion

This section presents the results obtained in the development and deployment of the CCN-51 cocoa fruit disease detection application, divided into two stages: i) interaction with the system, which explains how the application works, and ii) validation of the application, which analyses the results obtained once the planned tests were carried out when deploying the app.

Fig. 4 Execution flow of the mobile application and execution process

Table 1 Analysis of Results

Comparison of traditional methods && System		
	% Failure	% Hit
Traditional method	30–50	50%
Laboratory analysis	0.1–1%	99%
Application	0.4–20%	80–99%

5.1 Interaction with the System

The application consists of two modules, each one in charge of a specific function, such as:

Web Module Web module being the administrator module which is responsible for registering the various farms that access the system, as well as allowing to see the connected devices, the established locations, and a detail of the analyzed fruits.

Mobile Application It presents the user with a flow that allows the user to quickly adapt to the system, with (1) an introductory tutorial, (2) a guide to the various diseases of cocoa, (3) the camera section that allows the user to take a photo of the fruit. This photo will be loaded into the database for further analysis, and finally (4) the results display screen, which will show the photo of the fruit and the percentage of infection it has (see Fig. 4).

5.2 Validation

The application has a success rate of 80–99.5%. These results were obtained after subjecting the application to different environments and climate changes that could affect the quality of the photographs.

In addition, a comparison was made between the traditional methods in the area and the application, obtaining the following results (see Table 1).

As can be seen, the application has a high percentage almost comparable to the analysis carried out in a laboratory, with the difference of being a quick and more accessible means of prevention, as well as requiring less time for the detection of a pest or disease in cocoa fruits.

6 Conclusion and Future Work

Developing technological tools and methods for faster disease prevention in agricultural fields is an alternative that can enhance both the growth and quality of crops. Rapid disease prevention reduces the need for chemical treatments, thereby improving product quality and extending shelf life while preserving the natural taste of the crops.

Finally, the long-term objective of this project is to extend this application to the whole area of the city of "Ventanas", located in the province of "Los Rios", Ecuador.

References

1. S. Savary, A. Ficke, J.-N. Aubertot, and C. Hollier, "Crop losses due to diseases and their implications for global food production losses and food security," 2012.
2. B. Ney, M.-O. Bancal, P. Bancal, I. Bingham, J. Foulkes, D. Gouache, N. Paveley, and J. Smith, "Crop architecture and crop tolerance to fungal diseases and insect herbivory. mechanisms to limit crop losses," European Journal of Plant Pathology, vol. 135, no. 3, pp. 561–580, 2013.
3. F. N. Iandola, M. W. Moskewicz, K. Ashraf, S. Han, W. J. Dally, and K. Keutzer, "Squeezenet: Alexnet-level accuracy with 50x fewer parameters and <1mb model size," CoRR, vol. abs/1602.07360, 2016. [Online]. Available: http://arxiv.org/abs/1602.07360
4. F. N. Iandola, S. Han, M. W. Moskewicz, K. Ashraf, W. J. Dally, and K. Keutzer, "Squeezenet: Alexnet-level accuracy with 50x fewer parameters and¡ 0.5 mb model size," arXiv preprint arXiv:1602.07360, 2016.
5. S. Xie, R. Girshick, P. Dollar, Z. Tu, and K. He, "Aggregated ´ residual transformations for deep neural networks," arXiv preprint arXiv:1611.05431, 2016.
6. K. He, X. Zhang, S. Ren, and J. Sun, "Deep residual learning for image recognition," in The IEEE Conference on Computer Vision and Pattern Recognition (CVPR), June 2016.
7. M. Lin, Q. Chen, and S. Yan, "Network in network," arXiv preprint arXiv:1312.4400, 2013.
8. C. Szegedy, W. Liu, Y. Jia, P. Sermanet, S. Reed, D. Anguelov, D. Erhan, V. Vanhoucke, and A. Rabinovich, "Going deeper with convolutions," in Proceedings of the IEEE conference on computer vision and pattern recognition, 2015, pp. 1–9.
9. K. Simonyan and A. Zisserman, "Very deep convolutional networks for large-scale image recognition," arXiv preprint arXiv:1409.1556, 2014.
10. M. D. Zeiler and R. Fergus, "Visualizing and understanding convolutional networks," in European conference on computer vision. Springer, 2014, pp. 818–833.
11. A. Krizhevsky, I. Sutskever, and G. E. Hinton, "Imagenet classification with deep convolutional neural networks," in Advances in neural information processing systems, 2012, pp. 1097–1105.
12. K. P. Prabhakaran Nair, "The Agronomy and Economy of Important Tree Crops of the Developing World," Agron. Econ. Important Tree Crop. Dev. World, 2010.
13. A. Krizhevsky, I. Sutskever, and G. E. Hinton, "ImageNet Classification with Deep Convolutional Neural Networks," Neural Inf. Process. Syst., 2012.
14. K. He, X. Zhang, S. Ren, and J. Sun, "Deep Residual Learning for Image Recognition," in 2016 IEEE Conference on Computer Vision and Pattern Recognition (CVPR), 2016, vol. 7, no. 3, pp. 770–778.
15. D. G. Lowe, "Distinctive image features from scale invariant keypoints," Int. J. Comput. Vis., vol. 60, pp. 91–11020042, 2004.

16. H. Bay, T. Tuytelaars, and L. Van Gool, "SURF: Speeded up robust features," Lect. Notes Comput. Sci. (including Subser. Lect. Notes Artif. Intell. Lect. Notes Bioinformatics), vol. 3951 LNCS, pp. 404–417, 2006.
17. R. Girshick, J. Donahue, T. Darrell, and J. Malik, "Rich Feature Hierarchies for Accurate Object Detection and Semantic Segmentation," in 2014 IEEE Conference on Computer Vision and Pattern Recognition, 2014, pp. 580–587.
18. Republica del CACAO, «Ecuador, La cuna del Cacao,» 2022. [En línea]. Available: https://republicadelcacao.com/es/blogs/news/ecuador-the-home-of-cacao.
19. J. V. N. M. Emma Vargas, «SOCIO-ECONOMIC IMPACT OF THE PRODUCTION AND MARKETING OF COCOA, » ECOCIENCIA, vol. 8, n° 1390-9320, 2021.
20. M. d. A. y. Ganaderia, «Cacao Híbrido CCN-51 cuenta con certificación de calidad,» 2022. [En línea]. Available: https://www.agricultura.gob.ec/cacao-hibrido-ccn-51-cuenta-con-certificacion-de-calidad/.
21. AGROBANCO, «Manejo de enfermedades y plagas en el cultivo de CACAO,» 2020. [En línea]. Available: www.agrobanco.com.
22. A. Calvo, «Agroptima,» 2019. [En línea]. Available: https://www.agroptima.com/es/blog/tecnologia-agricultura-beneficios/.

Target Detection Algorithm of Forward-Looking Sonar Based on Swin Transformer

Lingyu Wang, Xiaofang Zhang, Shucheng Li, Guocheng Gao, Jianjun Wang, and Qi Wang

1 Introduction

The target information obtained by target detection is one of the basic information needed by AUVs to complete underwater operations. On the basis of the research of target detection based on optical image, the research of target detection algorithm for forward-looking sonar image has also been developed. In recent years, with the rapid development of forward-looking sonar applications in target recognition and obstacle avoidance, AUVs equipped with forward-looking sonar have been able to conduct autonomous navigation and obstacle avoidance.

In 2007, Clark et al. [1] proposed and tested two methods for tracking multiple targets using forward-looking sonar images. Galceran et al. [2] proposed a method of target detection on forward-looking sonar images in 2012. In 2016, Valdenegro-Toro [3] used sliding Windows to obtain a large number of prediction boxes on the forward-looking sonar images and screened them with intersection ratio. Then, CNN was used to predict. This method achieved high accuracy, but there was a lot of false detection. In 2017, Kim et al. [4] proposed an object detection network based on an FLS image, which used Haar-like features to construct multiple weak classifiers and combined them to form a strong classifier. The weight coefficient of each weak classifier was obtained by AdaBoost algorithm. In the same year, Dos Santo et al. [5] used a series of image processing techniques to improve the quality of sonar images,

L. Wang · G. Gao · J. Wang · Q. Wang
Ocean University of China, Qingdao, Shandong, China
e-mail: 21200211215@stu.ouc.edu.cn; wangqi7757@stu.ouc.edu.cn

X. Zhang (✉)
Naval Submarine Academy, Qingdao, Shandong, China

S. Li
James Cook University, Singapore, Singapore

L. Meng (ed.), *International Conference on Cloud Computing and Computer Network*, Signals and Communication Technology,
https://doi.org/10.1007/978-3-031-47100-1_2

tested and evaluated the performance of three classifiers (support vector machine, K-nearest neighbor and random tree), and the K-nearest neighbor classifier got the best results. In 2018, Ye et al. [6] combined the fully convolutional network (FCN) and the Siam network to propose a new object-tracking algorithm. The FCN structure solves the problem of the different image sizes of the input forward-looking sonar and improves the tracking efficiency and accuracy. Jin and his team [7] proposed a method of underwater target detection through forward-looking sonar and deep convolutional neural networks (DCNNs) in 2019 and proposed a training method based on transfer learning, which improved the network performance. In the same year, Kvasi [8] et al. found a robust and reliable target detection network for detecting and tracking divers. In 2020, Zacchini et al. [9] used Mask R-3CNN (Mask Regions with Convolutional Neural Network features) to recognize and locate potential objects in FLS images. Although the accuracy of target detection is high, it does not meet the requirement of real-time detection.

With the continuous deepening of marine resource exploitation, the target detection algorithm carried by AUV in the execution of tasks also needs to be more advanced and achieve higher accuracy on the basis of satisfying real-time performance. It is of great practical significance to design an object detection algorithm that balances real-time and precision.

Combined with the data characteristics of forward-looking sonar images, an FLS image target detection algorithm based on Swin transformer [10] is proposed in this paper. In the framework of Swin transformer, the improved FPN structure (Advanced FPN), long jump connection structure, and Wing Loss function are introduced to obtain Swin_FLS, a forward-looking sonar image target detection network with both real-time and accuracy. And, experiments are carried out to verify its high efficiency.

2 Swin_FLS Model

2.1 Dataset Introduction

The forward-looking sonar image data set used for training of this subject was collected by Gemini 720i multi-beam forward-looking sonar at the dock of Qingdao Scientific Investigation Center of State Oceanic Administration. A total of three common underwater entanglements are included in the data set, which are fishing net, cloth strip, and plastic bag. The specific objects and the feedback images of the forward-looking sonar are shown in Fig. 1. In order to simulate the real underwater operation scene of AUV and obtain a more diverse data set, the azimuth angle of entanglements and the distance between them and sonar are constantly changed during data collection. The sonar images of tangles of different shapes were obtained.

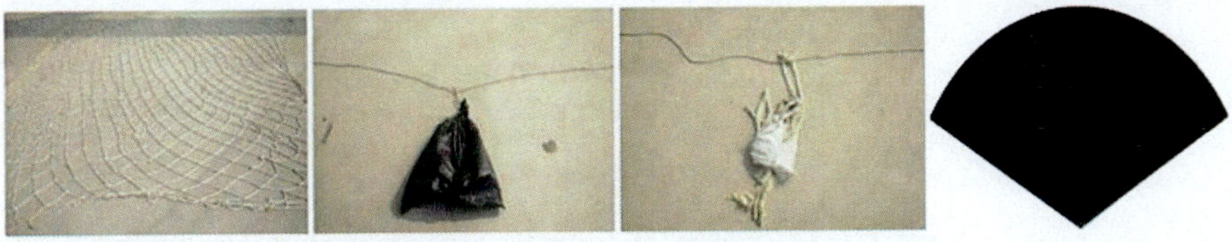

Fig. 1 (**a**) fishnet. (**b**) plastic bag. (**c**) cloth. (**d**) image of the forward-looking sonar

In this paper, the pixel-level image annotation tool LabelMe [11] developed by MIT was used to manually annotate the forward-looking sonar image and obtain the ground truth for the target detection algorithm. A total of 18,332 images were obtained through screening and manual labeling of forward-looking sonar images, in which the proportion of each category (fishing net, cloth strip, plastic bag) was about 1:1:1. Randomly select 80% of the images as the training set and the remaining 20% as the test set. In addition, about 20% of the images in the training set are randomly selected as the verification set. The training set data is responsible for training the target detection network. While verifying the effectiveness of the model, the verification set selects the model with good performance for storage, and the test set further tests the effectiveness and real-time performance of the storage model. More precisely, the data set consists of 10,995 training set images, 3667 verification set images, and 3670 test set images.

2.2 Network Description

The Swin_FLS model adopts two stages as a whole, with an anchor-based target detection method. The network model is composed of Backbone, FPN, and Head. In the backbone network part, Swin Transformer is used as the basis to extract features. Swin Transformer outputs four different stages, and its main function is to extract image features under different resolutions. On the basis of output feature maps of different sizes at the four stages of Swin Transformer, DenseCSP Module, a dense long jump connection module based on CSP structure, is added. Short connections are added between each stage in the network and each subsequent stage. Compared with other methods, Swin_FLS uses the Conv module to construct short connections, which can reduce the number of training parameters. More importantly, the features extracted by the CSP module are more sufficient, which makes the backbone network more stable when it is trained to the deep level after adding dense connections. For the FPN network module, a new FPN for feature fusion, Advanced-FPN, is proposed. Compared with the traditional FPN, top-down downsampling fusion is adopted, and residual connection of the middle two stages is added. The module carries out weighted feature fusion according to the output of the stage layer in the Swin feature extraction module. Finally, input the fused features into ReatinaHead to get the predicted results. Figure 2 shows the overall framework of Swin_FLS target detection network.

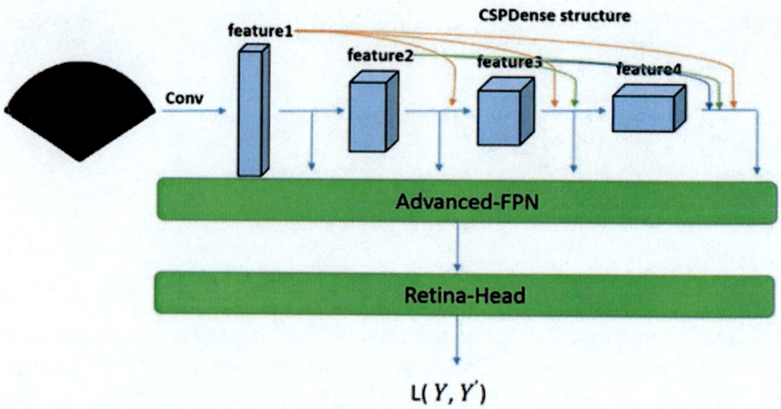

Fig. 2 Swin-FLS network structure

DenseCSP Module

Based on Swin Transformer, DenseCSP Module is added to the algorithm in this paper. In this paper, Densenet [12] neural network is referred to and its dense connection mechanism is referred to, that is, all layers are connected with each other, and each layer will accept all previous layers as its additional input. The advantage of this approach is that it can not only enable the lower stage to supplement rich texture features but avoid the features extracted by the lower network from being worthless. It can also avoid gradient disappearance and gradient explosion. Compared with ResNet, this structure can use supervised learning to train the weight of short connections between each feature layer in the process of model training, that is, which connections are more important to train. In addition, referring to the network structure of cspnet [13], the CSP structure is used in the connection mode between the feature layers, replacing the traditional simple convolution layer connection mode, which has the following advantages: improving the learning ability of CNN, keeping accuracy while lightweight, reducing computing bottlenecks and storage costs. This module is called DenseCSP Module because it adopts the dense connection mechanism of Densenet network and references the CSP structure connection module of CSPNet.

Advanced FPN

FPN [14] realizes the integration of features of different resolutions and dimensions of backbone network output, so as to construct high-level semantic features of various scales. However, in order to adapt to the characteristics of the forward-looking sonar data set used in this paper, we used an improved FPN structure (Advanced FPN), whose structure diagram is shown in Fig. 3.

For marine obstacle images detected by forward-looking sonar, the detection process pays more attention to the texture information rather than the semantic information, so the shallow features of the image are more important. If the common FPN structure is adopted and the output of each stage in the feature extraction

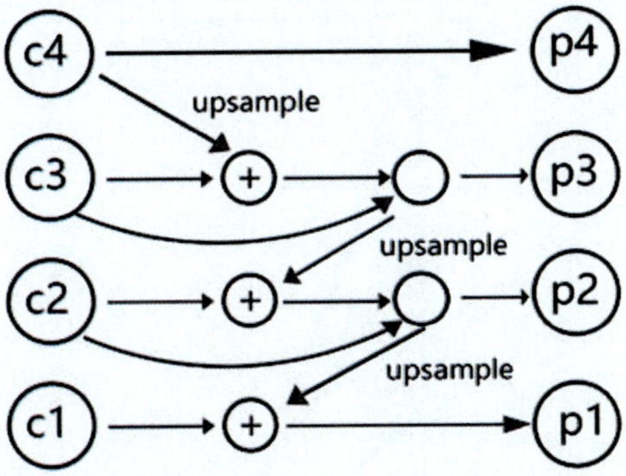

Fig. 3 Advanced FPN structure

network is processed in the same way, the shallow feature processing cannot be realized. In Advanced FPN, the shallow features (C2, C3) are connected by long jump, and the information is transferred downward after sampling up, so as to make full use of the shallow features and prevent the loss of information. For the deep feature C4 with a small contribution, the feature fusion is not carried out.

Wing Loss

This paper also uses the Loss combination of position regression Loss + classification Loss, and the Focal Loss used for classification loss. In the part of position loss, this paper puts forward the regression loss named Wing Loss. The following is the calculation formula of Wing Loss:

$$\text{wing}(x) = \begin{cases} wln\left(1 + \frac{|x|}{\epsilon}\right), & \text{if} |x| < w \\ |x| - C, & \text{otherwise} \end{cases} \tag{1}$$

Figure 4 shows the comparison of Wing Loss image with L1 and L2. Compared with the traditional L1, L2, and SmoothL1 losses, the Wing Loss convergence speed is faster, and the loss value can be controlled when the loss absolute value difference is greater than 1 and less than 1, so as to avoid the slower and slower convergence when the loss absolute value difference is less than 1 like L2 loss, and there is no gradient at zero like L1. And the convergence speed is faster than Smooth L1.

3 Analysis of Experimental Results

3.1 Ablation Study

This section conducts longitudinal comparison experiments on the model, analyzes the role of each module in the overall structure, and further verifies the rationality of the design of each module. In Table 1, six groups of ablation experiments were

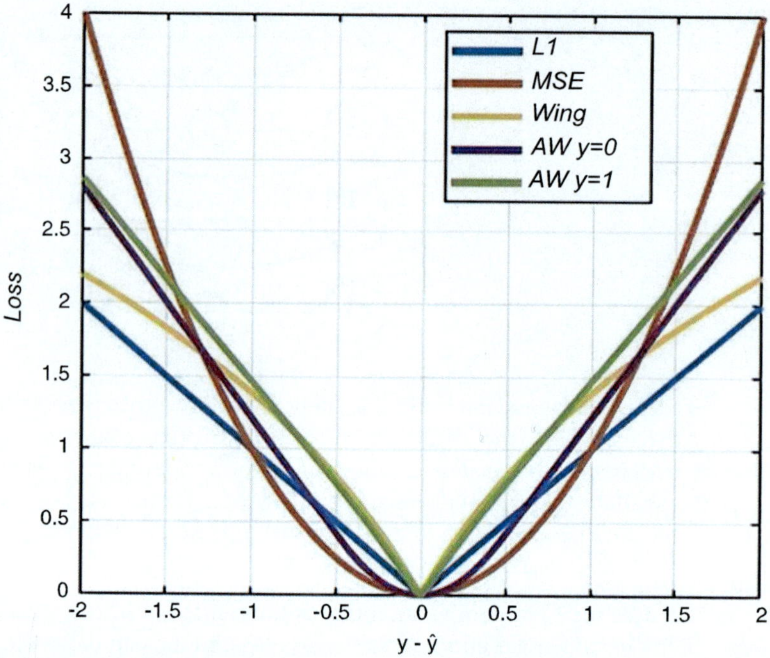

Fig. 4 Loss function comparison

Table 1 Ablation results

	DenseConv	DenseCSP	Advanced-FPN	Wing	mAP_{50}	mAP_{75}
Ex.1	×	×	×	×	81.8	74.3
Ex.2	√	×	×	×	75.4	68.2
Ex.3	×	√	×	×	83.5	78.5
Ex.4	×	×	√	×	78.9	70.7
Ex.5	×	×	×	√	80.5	77.6
Ex.6	√	×	√	√	**85.2**	**81.2**

conducted, corresponding to the experiments without any improved module, the experiments with only a single improvement, and the experiments with all improvements. As densely connected modules are special, the experiments constructed using convolution and those based on CSP module were also added to the ablation experiment table.

As can be seen from the information in Table 1, the three improvements have the following characteristics:

CSPDense-Swin: As for the index of Map(IOU = 0.5), CSPDense structure has a strong influence on the detection accuracy of the model. When convolution is used to build a dense connection module, the accuracy decreases by 5% on the basis of Base, but after the convolutional module is changed to CSP module, the accuracy rises to 83.5. It is also proved that using CSP structure to build a dense connection module can more fully extract the features of each Stage compared with convolutional module.

Advanced FPN: Single use of Wing Loss and Advanced FPN did not significantly improve the accuracy. However, when combined with DenseCSPSwin backbone network, Advanced FPN improved the accuracy by 2%. From this point of view, in the scene of forward sonar recognition in the field of target detection, the texture features extracted from the backbone network are crucial for the prediction of FPN and Head. When only Swin Transformer is used, the texture features in the image will disappear due to the increase of network depth. This is why using only Advanced FPNs does not yield good results.

Wing Loss: In the case of Map(IOU = 0.75), compared with the current situation and L1 Loss, Wing Loss makes the regression of target detection prediction box more accurate. When the IOU threshold is increased, the drop range of the map is less than that of L1, which indicates that the wing loss can be dynamically adjusted according to the training state.

3.2 Comparative Experiment

In this section, the target detection performance of Swin_FLS is compared with that of classical neural network to test whether Swin_FLS has superiority in underwater obstacle detection. The experimental results are shown in Table 2.

As can be seen from Table 2, mAP index of detection accuracy of the proposed method reaches 85.2, which is higher than that of classical networks like Faster RCNN, SSD, and RetinaNet [15–17] and the newly launched YOLOV5 network in recent years, and meets the requirement of high precision when AUV performs tasks.

3.3 Real-Time Experiment

As the final neural network model is to be applied to the real-time detection of obstacles and deployed on the embedded device (Nvidia Jetson Xavier) installed on the AUV, real-time performance and embeddability are also critical. In practical

Table 2 Comparative result

Method	mAP	$AP_{fishnet}$	$AP_{plasticbag}$	AP_{cloth}
Faster RCNN	56.4	41.2	68.9	59.1
SSD	60.6	44.7	70.1	67.0
RetinaNet	74.3	69.2	80.3	73.4
Dert	72.6	63.1	81.3	75.8
Yolov5-S	75.2	63.1	84.8	77.7
Swin_FLS	**85.2**	**74.4**	**90.6**	**90.7**

Table 3 Real-time comparison

Method	Picture size	Operation time (ms)	FLOPS(B)	FPS	Model size (Mb)
Faster RCNN	(640,640)	800	310	1.2	138.5
SSD	(640,640)	45	12	21.9	22
RetinaNet	(640,640)	198	74	5.1	46
Dert	(640,640)	531	201	1.9	159
Yolov5-S	(640,640)	61	16	16.2	7.2
Swin_FLS	(640,640)	70	21	14.1	43

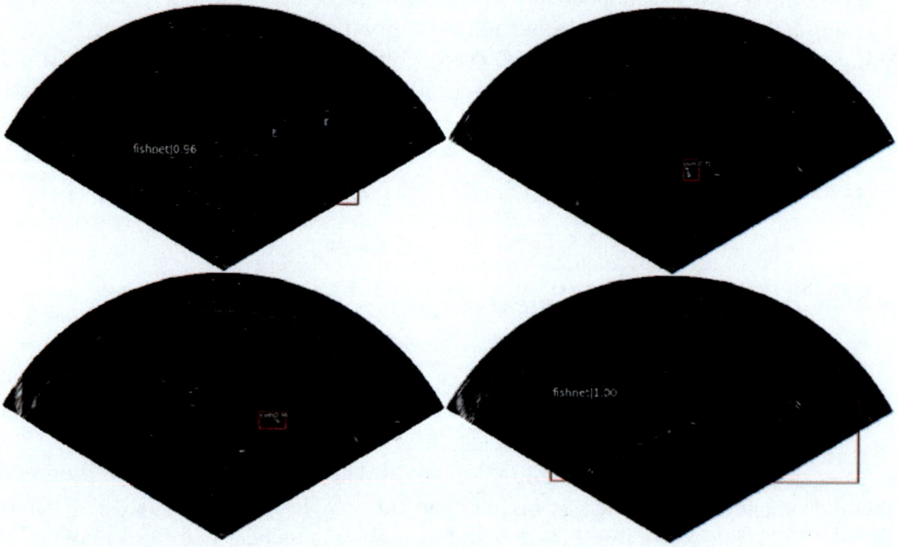

Fig. 5 Test result graph

application, the real-time performance reaches 10 frames/second or more, which can meet the needs of obstacle avoidance. This section will test whether it can meet the requirements of practical application by comparing it with some classical algorithms and the network before improvement. The experimental results are shown in Table 3. It can be seen that the algorithm in this paper can meet the requirements of real-time detection.

3.4 Actual Target Detection Effect

Figure 5 shows the final target detection results of Swin_FLS model on the actual forward-looking sonar image.

4 Conclusion

In this paper, a deep learning-based forward sonar image target detection algorithm, Swin_FLS, is proposed. The attention mechanism is used to extract sonar image features, and the CSPDense structure is used to strengthen the association between feature images, enrich the texture information of the deep feature layer, and focus on processing the middle layer feature information. Wing loss is used to train the model, which effectively prevents the gradient explosion and speeds up the convergence of the model. Finally, this paper verifies the effectiveness and real-time performance of the proposed algorithm through a large number of experiments and achieves good results. However, limited by the number of data sets, there is still a gap with the actual ocean target detection. Due to the variable forms of obstacles in the ocean and the complex and changeable environment, the detection stability of Swin_FLS in different marine environments needs to be further improved.

References

1. Clark D, Ruiz I T, Petillot Y, et al. Particle PHD filter multiple target tracking in sonar image [J]. IEEE Transactions on Aerospace and Electronic Systems, 2007, 43(1):409–416.
2. Galceran E., Djapic V., Carreras M., Williams D.P. A real-time underwater object detection algorithm for multi-beam forward looking sonar. IFAC Proceedings Volumes, 2012, 45 (5): 306–311.
3. Valdenegro-Toro M. Submerged Marine Debris Detection with Autonomous Underwater Vehicles[C]. IEEE International Conference on Robotics and Automation for Humanitarian Applications, Kollam, India, 2016.
4. Kim B, Yu S C. Imaging sonar based real-time underwater object detection utilizing AdaBoost method[C]. IEEE Underwater Technology, Busan, South Korea, 2017.
5. Matheus D S, Pedro R, Pedro N, et al. Object Classification in Semi Structured Enviroment Using Forward-Looking Sonar[J]. Sensors, 2017, 17(10).
6. Ye X, Sun Y, Li C. FCN and Siamese Network for Small Target Tracking in Forward-looking Sonar Images[C]. MTS/IEEE OCEANS, Charleston, SC, USA, 2018
7. Jin L, Liang H, Yang C. Accurate Underwater ATR in Forward-Looking Sonar Imagery Using Deep Convolutional Neural Networks[J]. IEEE Access, 2019, 7:125522–125531
8. Kvasic I, Miskovic N, Vukic Z. Convolutional Neural Network Architectures for Sonar-Based Diver Detection and Tracking[C]. IEEE OCEANS, Marseille, France, 2019.
9. Zacchini L, Franchi M, Manzari V, Pagliai M, Ridolfi A. Forward-Looking Sonar CNN-based Automatic Target Recognition: an experimental campaign with FeelHippo AUV[C]. IEEE Conference on Autonomous Underwater Vehicles Symposium (AUV), St. Johns, NL, Canada, 2020.
10. A. Yueyuan and W. Hong, "Swin Transformer Combined with Convolutional Encoder For Cephalometric Landmarks Detection," 2021 18th International Computer Conference on Wavelet Active Media Technology and Information Processing (ICCWAMTIP), Chengdu, China, 2021, pp. 184–187, doi: https://doi.org/10.1109/ICCWAMTIP53232.2021.9674147.
11. Russell B C, Torralba A, Murphy K P, et al. LabelMe: a database and web-based tool for image annotation[J]. International journal of computer vision, 2008, 77(1–3): 157–173

12. H. Dong, L. Zhang and B. Zou, "Densely Connected Convolutional Neural Network Based Polarimetric SAR Image Classification," IGARSS 2019–2019 IEEE International Geoscience and Remote Sensing Symposium, Yokohama, Japan, 2019, pp. 3764–3767, doi: https://doi.org/10.1109/IGARSS.2019.8900292.
13. C. -Y. Wang, H. -Y. Mark Liao, Y. -H. Wu, P. -Y. Chen, J. -W. Hsieh and I. -H. Yeh, "CSPNet: A New Backbone that can Enhance Learning Capability of CNN," 2020 IEEE/CVF Conference on Computer Vision and Pattern Recognition Workshops (CVPRW), Seattle, WA, USA, 2020, pp. 1571–1580, doi: https://doi.org/10.1109/CVPRW50498.2020.00203.
14. Lin T.Y, Dollar P, Girshick R, He K, Hariharan B, Belongie S. Feature Pyramid Networks for Object Detection[C]. IEEE Conference on Computer Vision and Pattern Recognition, Honolulu, Hawaii, 2017.
15. Ren S, He, K, Girshick R, Sun J. Faster R-CNN: Towards Real-Time Object Detection with Region Proposal Networks[J]. IEEE Transactions on Pattern Analysis & Machine Intelligence 2015, 39(6), 1137–1149.
16. Liu W, Anguelov D, Erhan D, Szegedy C, Reed S, Fu C, Berg A. C. SSD: Single Shot MultiBox Detector[C]. European Conference on Computer Vision, Amsterdam, The Netherlands, 2016.
17. T. -Y. Lin, P. Goyal, R. Girshick, K. He and P. Dollár, "Focal Loss for Dense Object Detection," 2017 IEEE International Conference on Computer Vision (ICCV), Venice, Italy, 2017, pp. 2999–3007, doi: https://doi.org/10.1109/ICCV.2017.324.

An Optimization Strategy for Efficient Facial Landmark Detection Based on Improved Pixel-in-Pixel Net Model

Renhao Li, Yanan Yu, and Guanghua Yin

1 Introduction

With the development of science and technology, the technology of efficient facial landmark detection is widely used in various fields, such as driverless driving, facial beautification technology, and facial expression analysis. Facial retouching technology is based on the accurate analysis of face shape, eye shape, nose shape, etc. It can modify and process the specific position on the face to manage some entertainment functions like facial beautification technology and improve facial editing algorithms. Facial expression analysis based on key points that analyze human facial expressions can be used in interactive entertainment, behavior prediction, and other scenarios. In algorithms such as facial pose alignment and facial recognition, it is necessary to detect feature points on the face to achieve pose alignment and improve the accuracy of the model. In modern life, people have been using facial detection technology more often than ever, so more accurate and efficient facial landmark detection is needed.

Facial landmark detection refers to locating the key points of the face in a given face image, including points in the eyebrows, eyes, nose, mouth, and facial contour areas, which are affected by factors such as poses and obstructions. Facial landmark detection is a challenging task, because each of its key points is a dense point selected from a face, and finally connected together to lines to form eyebrows, eyes, nose, mouth, and facial contour. Facial landmark detection is the core content in the facial landmark detection technology. Through key points, each part of the face can be composed and connected into a single area. The coordinates of key points affect the size and position of each area.

R. Li · Y. Yu (✉) · G. Yin
School of Information Technology and Engineering, Tianjin University of Technology and Education, Tianjin, China

L. Meng (ed.), *International Conference on Cloud Computing and Computer Network*, Signals and Communication Technology,
https://doi.org/10.1007/978-3-031-47100-1_3

There is an issue with facial landmark detection: key point position detection can be inaccurate. For example, the key points of the face could drift sometimes, and there is an error between the coarse positioning and fine positioning of the key points of the face, which makes information of key points on the face incorrect and failure in recognizing information correctly. In response to the aforementioned issues, this paper focuses on the drift of facial key point prediction results, the error between facial coarse positioning and fine positioning, etc. Studying thoroughly the facial information extraction algorithm based on facial landmark detection, improving the pixel-in-pixel network model, and choosing a more advanced network to further improve the accuracy of facial landmark detection and reduce the failure rate of facial landmark detection. The improvement of the accuracy of facial landmark detection plays a vital role in selecting facial key points; it avoids huge errors where marking facial key points and makes the information of facial key points more accurate. The deviation of the coordinate position may affect each node area of the face and make a greater deviation, such as the meaningless increase or loss of facial contour. Through the normalized average error and the area under the curve, the two major indicators of facial landmark detection, we can judge whether the selected network can improve the accuracy and reduce the loss rate or not. We can select the network with the smallest standardized average error and the largest area under the curve after comparing different networks.

This paper aims at the problem that the normalized average error of the pixel-in-pixel network is too large and the area under the curve is too small; the improved MobileNetV2_re is used to effectively reduce the loss of Pixel-in-pixel. The improved model uses three ghost bottleneck layers to replace the original fifth block in MobileNetV2. The inverted residual module has a better performance in reducing the standardized average error and increasing the area under the curve compared with other networks.

2 Related Work

2.1 PIPNet

Facial landmark detection is a very important pre-task in applications related to facial recognition, and face applications are real-time and uncertain, which makes the facial landmark model require not only high accuracy but also high-speed computing and a strong ability of anti-interference. However, many methods usually cannot meet all these requirements at the same time. For example, the method based on heat map regression can achieve high accuracy, but it involves a large computational load and it is sensitive to the cluster samples. On the contrary, the one-stage method based on coordinate regression is fast and robust, but the prediction accuracy is not high. Although coordinate regression can be extended to a multi-stage approach to improve accuracy, its computational cost also increases. In addition, the cross-domain generalization ability of facial landmark models is also a challenge. In order to obtain a facial key point model that is suitable for practical applications, Haibo Jin et al. proposed a pixel-in-pixel Net (PIPNet) model [1] to deal with the above situations, which mainly includes three important modules. The first is a novel detection head called pixel-in-pixel regression (PIP regression). This method decomposes the key point positioning task into heat map regression based on low-resolution feature maps and coordinates regression on local feature maps, so that the model still has high accuracy without relying on high-resolution feature maps, thus the computational load is reduced. In addition, a neighbor regression block is additionally designed for the detection head. By training each key point to locate its neighbor key points according to its own position, which can help to obtain the shape constraints of the local area can be obtained during prediction, thereby improving the robustness of the model. Finally, a semi-supervised learning method based on self-training is proposed to make full use of a large number of unlabeled samples in different scenarios. When this method estimates pseudo-labels for unlabeled samples, it starts first with a simple task, and then gradually increases the difficulty of the task in subsequent iterations until it becomes a standard self-training task, which effectively alleviates the noise problem of the standard self-training method. The facial landmark detector PIPNet has managed to be fast, accurate, and stable, which is especially suitable for practical applications. According to the experimental results on six mainstream datasets, the proposed PIPNet has achieved leading prediction performance in three datasets. In addition, the model has reached the leading level in the balance of accuracy and speed in experiments across datasets.

PIPNet can be trained under two settings: (1) supervised learning; and (2) generalizable semi-supervised learning (GSSL). Using GSSL [2], PIPNet has better cross-domain generalization performance by utilizing a large amount of unlabeled data across domains.

The NME results of each network structures on the WFLW full test set are shown in Fig. 1.

This figure proves that compared with MapNet [3] and CoordNet [4], the normalized average error of PIPNet is lower, and the error is smaller. PIPNet is more accurate through the NME result experiment carried out under different frame numbers on CPU and GPU.

The comparison of different detection heads of PIPNet is shown in Fig. 2.

Figure 2 shows different detection heads: (a) Coordinate Regression; (b) Heatmap Regression; (c) PIP regression; (d) PIP regression + NRM.

2.2 *Resnet*

Resnet was proposed by He Kaiming when he worked at Microsoft Research Asia in 2015. Resnet won the first place in ImageNet classification, detection, positioning, and COCO detection and segmentation in 2015.

Fig. 1 Test results in NME under each Net structure. (**a**) CPU. (**b**) GPU

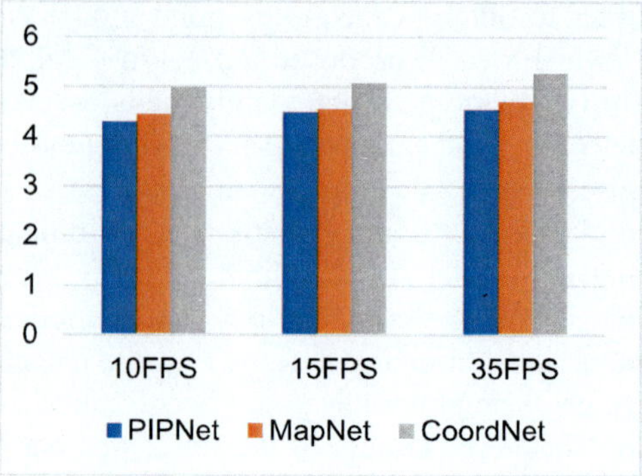

(a)

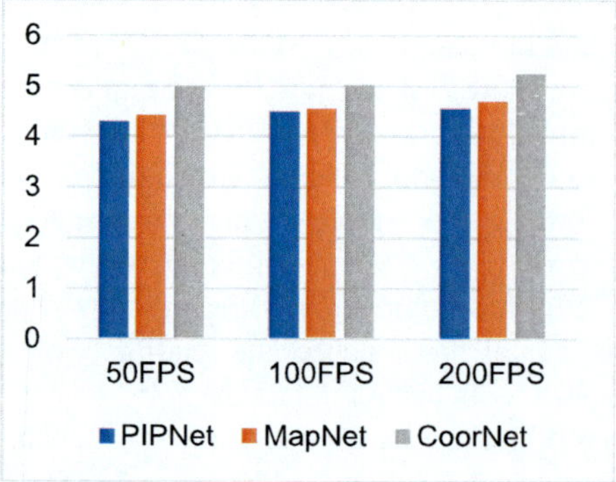

(b)

Now supposed $x = 5$, $F(x) = 5.1$, if it is a non-residual structure, then the network mapping is: $F(5)' = 5.1$, if it is a residual structure, the network mapping is: $F(5) + 0.1 = 5.2$.

Both F' and F here represent the network parameter mapping, and the mapping is more sensitive to the change of the output after introducing the residual. For example, the output of the mapping F' increases to 5.2 from 5.1, which is a $1/51 = 2\%$ increase and when it comes to the residual structure, it is increased from 5.1 to 5.2 and the mapping F is increased from 0.1 to 0.2, which is a 100% increase. Obviously, the output change of the latter has a greater effect on the adjustment of the weight, so the effect is better. Subsequent experiments also prove this hypothesis, and the residual network is a better network to train compared with the plain network. Therefore, ResNet solves the problem of how to train the network better.

The basis of the Resnet network [5] is based on the residual block. The differences between ResNet18, ResNet50, and ResNet101 are shown in Table 1. The basic meaning of ResNet18 [6] is that the basic framework of the network is ResNet, and the network depth is 18 layers.

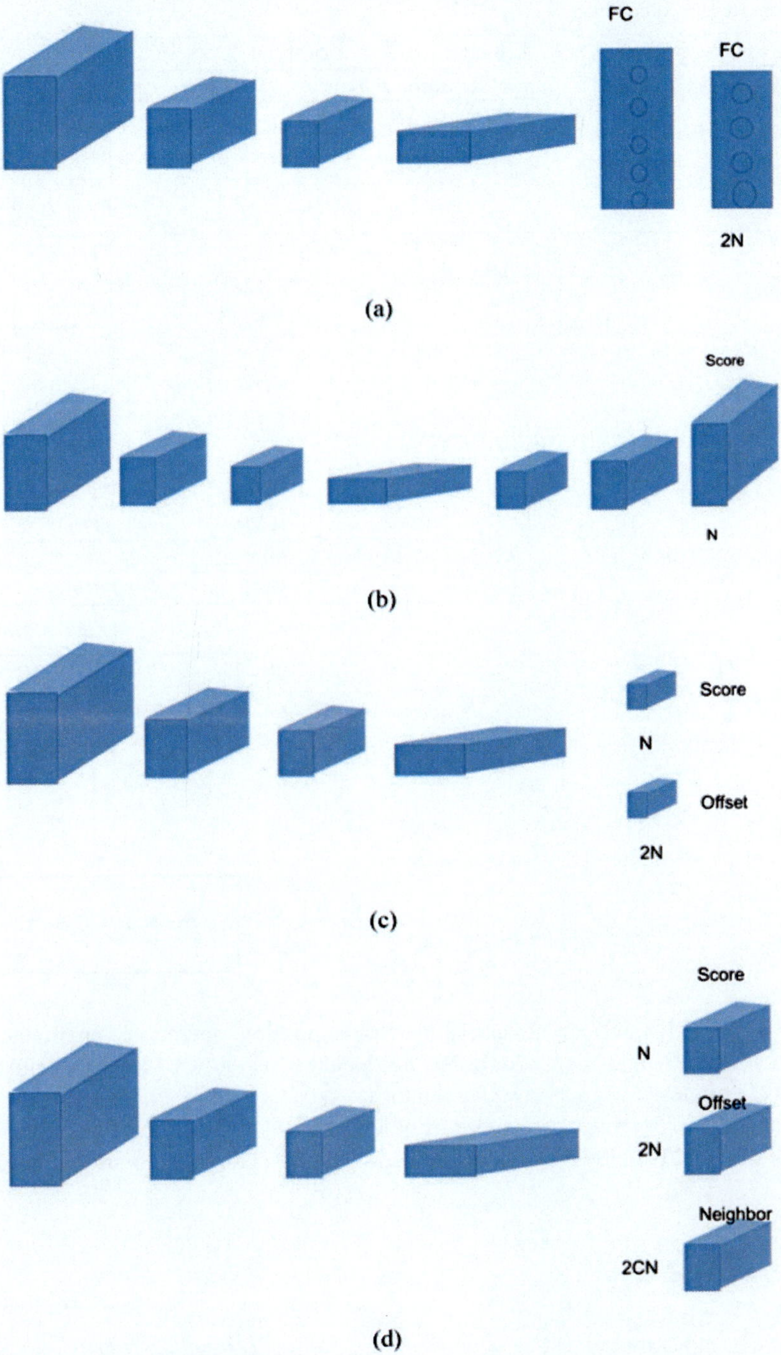

Fig. 2 Comparison of different detection heads. (**a**) coordinate regression. (**b**) heatmap regression. (**c**) PIP regression. (**d**) PIP regression + NRM

Table 1 Resnet comparison

Layer name	Output size	18-layer	50-layer	101-layer
Conv1	112*112	7*7,64,stride 2		
Conv2_x	56*56	3*3 max pool,stride 2		
		$\begin{bmatrix}3*3, 64\\3*3, 64\end{bmatrix}*2$	$\begin{bmatrix}1*1, 64\\3*3, 64\\1*1, 256\end{bmatrix}*3$	$\begin{bmatrix}1*1, 64\\3*3, 64\\1*1, 256\end{bmatrix}*3$
Conv3_x	28*28	$\begin{bmatrix}3*3, 128\\3*3, 128\end{bmatrix}*2$	$\begin{bmatrix}1*1, 128\\3*3, 128\\1*1, 512\end{bmatrix}*4$	$\begin{bmatrix}1*1, 128\\3*3, 128\\1*1, 512\end{bmatrix}*4$
Conv4_x	14*14	$\begin{bmatrix}3*3, 256\\3*3, 256\end{bmatrix}*2$	$\begin{bmatrix}1*1, 256\\3*3, 256\\1*1, 1024\end{bmatrix}*6$	$\begin{bmatrix}1*1, 256\\3*3, 256\\1*1, 1024\end{bmatrix}*23$
Conv5_x	7*7	$\begin{bmatrix}3*3, 512\\3*3, 512\end{bmatrix}*2$	$\begin{bmatrix}1*1, 512\\3*3, 512\\1*1, 2048\end{bmatrix}*3$	$\begin{bmatrix}1*1, 512\\3*3, 512\\1*1, 2048\end{bmatrix}*3$
	1*1	Average pool,1000-d fc,softmax		
FLOPs		1.8*109	3.8*109	7.6*109

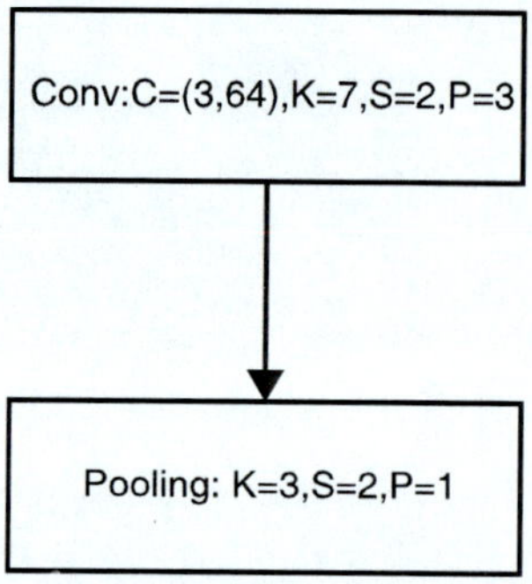

Fig. 3 Image pooling

Resnet50 [7] based on Resnet18 performs pooling operations on images, as shown in Fig. 3, and then adds three bottlenecks, as shown in Fig. 4, and adds various bottlenecks to each layer by analogy.

Resnet101 [8] has 17 more bottlenecks in conv4_x than Resnet50. 17*3 + 50 = 101. It copies 17 times and adds them to conv4_x, as shown in Fig. 5.

2.3 *MobileNetV2*

Compared with MobileNetV1 [9], the innovation of MobileNetV2 [10] uses many inverted residual blocks and linear bottleneck layers to extract image features. Experiments show that using linear layers can effectively avoid nonlinear layers

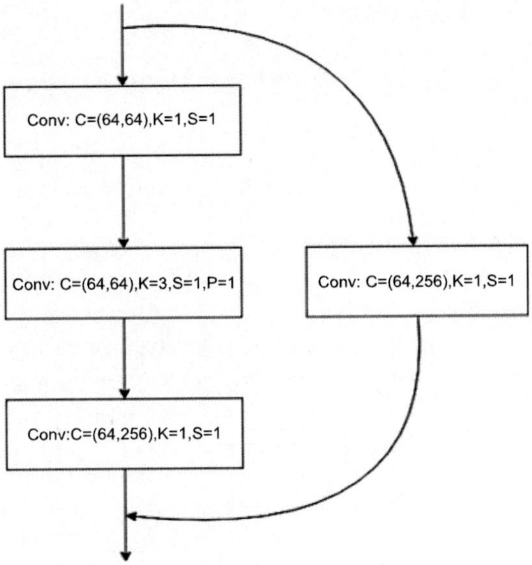

Fig. 4 Bottleneck of Resnet50

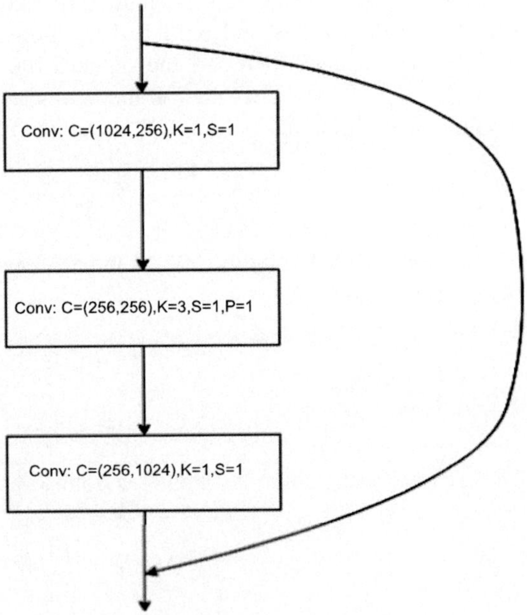

Fig. 5 Bottleneck of Resnet101

from destroying feature information. The inverted residual structure uses point-by-point convolution to increase the dimension firstly, then uses depth convolution, and finally uses point-by-point convolution to reduce the dimension. This can reduce the parameters and calculations. Also, it can extract useful features better.

3 Algorithms

3.1 NME (Normalized Mean Error)

The average error [11] refers to the arithmetic mean of the random errors of all measured values measured in equal precision measurement, defined as:

$$\eta = \frac{1}{n}\sum\nolimits_{i=1}^{n} |x_i - \bar{x}| \tag{1}$$

Among them, η represents the average error, $\bar{x}$ and x_i represent the predicted value and the real value of the feature points, n represents the number of times of measurement, and the average error η represents the average value of $|x_i - \bar{x}|$.

The normalized mean error [12] of a single facial image is defined as:

$$NME = \frac{1}{N}\sum\nolimits_{j=1}^{N} \frac{\| p_j - \mu_j \|_2}{d} \times 100\% \tag{2}$$

Among them, N represents the number of pictures, and p_j and μ_j represent the true value of feature points and predicted position. According to different subscripts, set d as the distance between the outer corners of the eyes, when we calculate $NME_{\text{inter - ocular}}$, set d as the geometric mean of the width W and height H of the ground-truth bounding box of the supplied data when we calculate NME_{box}:

$$\sqrt{W_{\text{box}} \cdot H_{\text{box}}} \tag{3}$$

If ground-truth bounding box is not provided, the tight bounding box of the feature points can replace it.

3.2 AUC

The area under the curve AUC [13] is defined as:

$$AUC_a = \int_0^a f(e)de \tag{4}$$

Among them, a represents the upper limit taken, e represents the normalized error, $f(e)$ represents the cumulative error distribution function, and AUC_{box} sets d as the width and height of the image bounding box. To calculate AUC, the ROC curve should be drawn. AUC is evaluated by calculating the area under the curve. This method does not change greatly due to the large error of a single point like the square error.

3.3 MobileNetV2 Network Integrated into Ghost Module

In order to solve the redundancy of deep neural network feature maps, Han et al. proposed the ghost module [14], which generates more feature maps by linear operations, which not only ensures the accuracy of the model but also reduces the parameters of the model. Assuming that a feature map of n channels is generated, the scaling ratio is s, the size of the convolution kernel is k, and the number of channels of the input data is c, the formula we use to calculate the parameter compression ratio of ghost module is shown below (5):

$$r_c = \frac{n \times c \times k \times k}{\frac{n}{s} \times k \times k + (s-1) \times \frac{n}{s} \times k \times k} \tag{5}$$

The parameter quantity of the network can be reduced to the 1/s of original size. In the improved model, three ghost bottleneck layers are used to replace the original fifth inverted residual module in MobileNetV2. The ghost bottleneck layer consists of two ghost modules. The first module is used as an expansion layer to increase the number of channels. The second module reduces the number of channels. Then connect the input and output of the two ghost modules. The Ghost bottleneck with a step size of two layers is inserted into DW convolutions [15], which reduces the parameter scale again. The ghost bottleneck layer structure is shown in Fig. 6, and the improved MobileNetV2_re network structure is shown in Fig. 7.

4 Experiments and Results

4.1 Experimental Data

The 300 W dataset [16] is a common facial alignment dataset from real-life scenarios whose indexes are widely compared with indexes in papers to publish. The 300 W dataset was released in 2013. It contains 300 pictures taken indoors and 300 pictures taken outdoors. In this database, human facial expressions in the pictures change significantly, the lighting conditions in the pictures, and the human body. Pose, facial occlusion, and facial size. This dataset uses 68 key points methods; this method produces satisfactory results both in this dataset and others.

The WFLW dataset [17] is mainly a facial key point localization dataset, which is mainly used to train facial alignment algorithms. WFLW was released in 2018 and contains 10,000 face images, including 7500 training images and 2500 test images, using 98 key point annotation methods. In addition to the key points, there are also annotations of information such as face occlusion, human posture, lighting condition and facial expressions.

The COFW dataset [18] is primarily to present faces in real life, so the data in the COFW dataset has a more obvious improvement in occlusion and posture. The COFW dataset contains a total of 1007 images collected from various sources, and all images are annotated with 29 key points.

Fig. 6 Ghost bottleneck layer. (a) stride = 1 bottleneck. (b) stride=2 bottleneck

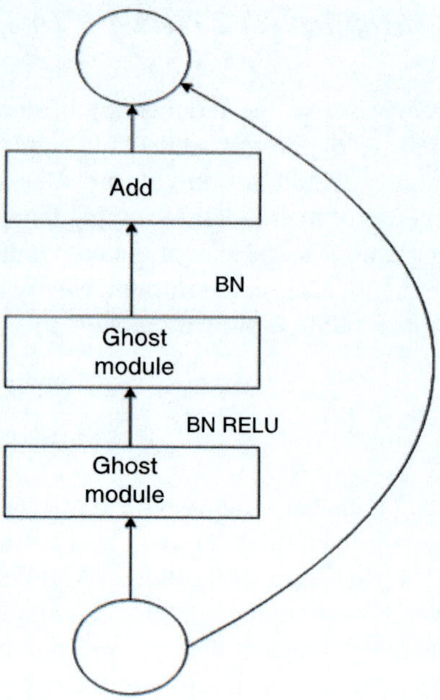

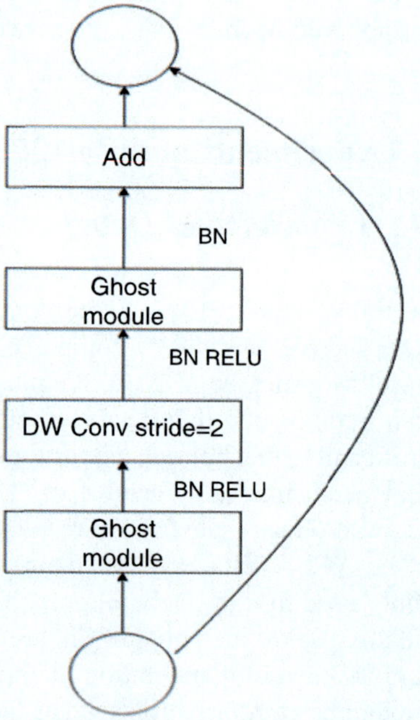

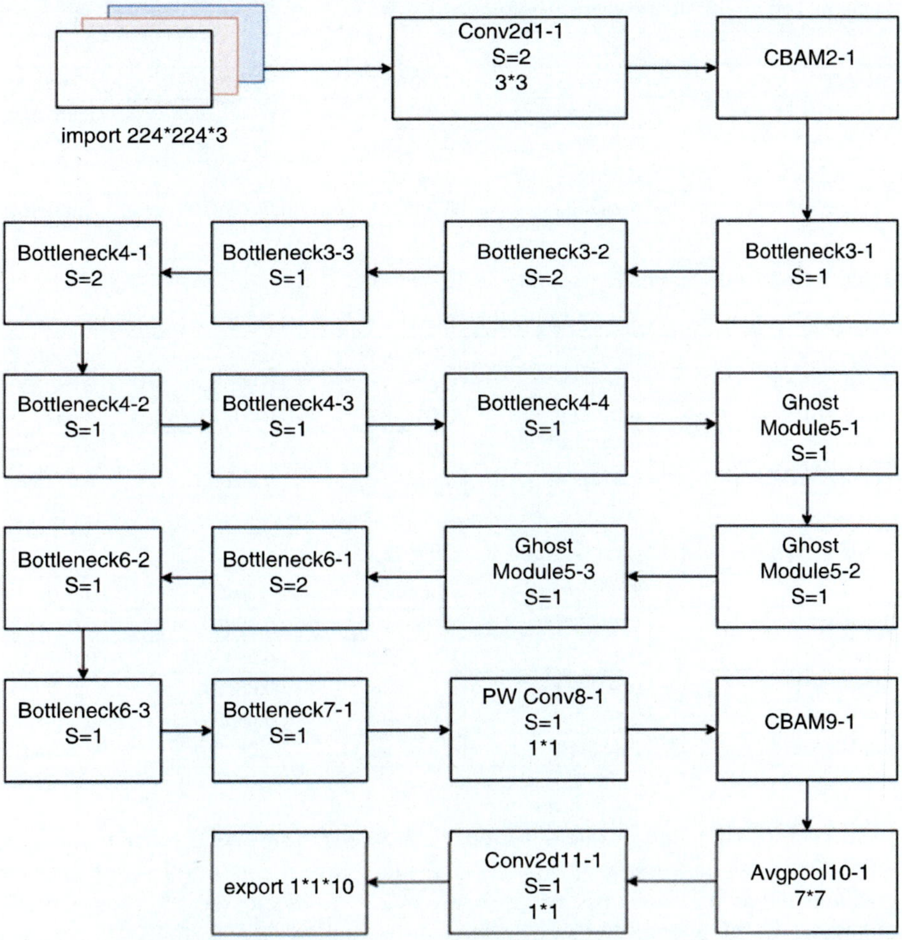

Fig. 7 Improved MobileNetV2_re network structure

4.2 Experimental Results

The test results in the dataset 300 W, WFLW, and COFW are shown in Tables 2, 3, and 4, respectively.

In Table 2, the total loss rate of MobileNetV2_re is 0.263403, the map loss rate is 0.035245, the x loss rate is 0.059319, and the y loss rate is 0.051514. It can be concluded that the total loss [19] of MobileNetV2_re is decreased by about 18.8% compared to MobileNetV2, the map loss decreased by about 29.5%, the x loss decreased by about 11.2%, the y loss decreased by about 20.1% when they are run on the 300 W dataset.

Table 2 Test results of different networks in the 300 W

	Total loss	Map loss	X loss	Y loss
Resnet18	0.316948	0.046745	0.067189	0.065520
Resnet50	0.305177	0.042914	0.065746	0.063219
Resnet101	0.272075	0.036727	0.062656	0.053627
MobileNetV2	0.324239	0.049994	0.066801	0.064442
MobileNetV2_re	**0.263403**	**0.035245**	**0.059319**	**0.051514**

Table 3 Test results of different networks in the WFLW

	Total loss	Map loss	X loss	Y loss
Resnet18	0.317146	0.042443	0.076424	0.063678
Resnet50	0.305304	0.041474	0.070267	0.059009
Resnet101	0.271669	0.034387	0.063706	0.052375
MobileNetV2	0.325129	0.046067	0.071819	0.062786
MobileNetV2_re	**0.271226**	**0.033928**	**0.062422**	**0.051642**

Table 4 Test results of different networks in COFW

	Total loss	Map loss	X loss	Y loss
Resnet18	0.424433	0.054710	0.090691	0.074638
Resnet50	0.421579	0.053523	0.089923	0.070321
Resnet101	0.362993	0.038548	0.082990	0.062784
MobileNetV2	0.457424	0.062794	0.093097	0.074818
MobileNetV2_re	**0.361610**	**0.036888**	**0.082901**	**0.054657**

In Table 3, the total loss rate of MobileNetV2_re is 0.271226, the map loss rate is 0.033928, the x loss rate is 0.062422, and the y loss rate is 0.051642. It can be concluded that the total loss of MobileNetV2_re is decreased by about 16.6% compared to MobileNetV2,the map loss decreased by about 26.4%, the x loss decreased by about 13.1%, the y loss decreased by about 17.7% when they are run on the WFLW dataset.

In Table 4, the total loss rate of MobileNetV2_re is 0.361610, the map loss rate is 0.036888, the x loss rate is 0.082901, and the y loss rate is 0.054657. It can be concluded that the total loss of MobileNetV2_re is decreased by about 20.9% compared to MobileNetV2,the map loss decreased by about 41.3%, the x loss decreased by about 11.0%, the y loss decreased by about 26.9% when they are run on the COFW dataset.

According to Tables 2, 3 and 4, it is obvious that compared with MobileNetV2, resnet18, resnet50 and resnet101, MobileNetV2_re has the least loss rate in the total loss, map loss, x loss, and y loss when run on the 300 W,WFLW, and COFW datasets. It can be concluded that MobileNetV2_re will be a better training network.

Under different datasets, the test results of NME and AUC in different networks are shown in Tables 5 and 6, respectively.

Table 5 NME under different networks in different datasets

	300 W	WFLW	COFW
MobileNetV2	0.035366	0.048849	0.035874
Resnet18	0.034767	0.047248	0.035721
Resnet50	0.033522	0.045600	0.032970
Resnet101	0.032403	0.044933	0.031819
MobileNetV2_re	**0.032036**	**0.043074**	**0.030828**

Table 6 AUC under different networks in different datasets

	300 W	WFLW	COFW
MobileNetV2	0.647605	0.542746	0.641741
Resnet18	0.652697	0.551308	0.643077
Resnet50	0.664980	0.563161	0.670278
Resnet101	0.676153	0.573906	0.681836
MobileNetV2_re	**0.679719**	**0.584435**	**0.691702**

From Tables 5 and 6, it is clear that the NME of MobileNetV2_re is reduced by about 9.42% compared with the MobileNetV2, and the AUC of MobileNetV2_re is increased by about 4.96% compared with MobileNetV2 when run on the 300 W dataset; NME of MobileNetV2_re reduced by about 11.82% comparing with MobileNetV2, and the AUC is increased by about 7.68%, when run on the WFLW dataset; NME of MobileNetV2_re is reduced by about 14.07% comparing with MobileNetV2, and the AUC is increased by about 7.52%, when run on the COFW data set. In addition, compared with MobileNetV2, resnet18, resnet50, and resnet101, MobileNetV2_re shows the least NME and the largest AUC on the 300 W, WFLW, and COFW databases, which also proves that MobileNetV2_re can become a better training network.

5 Conclusion

Aiming at the problems of high loss rate, large normalized average error, and small area under the curve in the original MobileNetV2 network, this paper proposes an optimization strategy for facial landmark detection based on the improved pixel-in-pixel network model which adds ghost module to the original MobileNetV2 network module. Then train a new network on MobileNetV2_re. Comparing with MobileNetV2, the normalized mean error of MobileNetV2_re network decreased by about 14.07%, the area under the curve increased by about 7.52%; comparing with resnet18, the normalized mean error of MobileNetV2_re network decreased by about 13.69%, the area under the curve increased by about 7.56%; comparing with resnet50, the normalized mean error of MobileNetV2_re network decreased by about 6.49%, the area under the curve increased by about 3.19%; comparing with resnet101, the normalized mean error of MobileNetV2_re network decreased by about 3.11%, the area under the curve increased by about 1.45%. This method can improve the detection accuracy in the facial landmark detection task.

References

1. Jin H, Liao S, Shao L. Pixel-In-Pixel Net: Towards Efficient Facial Landmark Detection in the Wild[J]. 2020.
2. Kipf T N, Welling M. Semi-Supervised Classification with Graph Convolutional Networks [J]. 2016.
3. Li Q, Wang Y, Wang Y, et al. HDMapNet: An Online HD Map Construction and Evaluation Framework[J]. 2021.
4. Hou Q, Zhou D, Feng J. Coordinate Attention for Efficient Mobile Network Design[J]. 2021.
5. He K, Zhang X, Ren S, et al. Deep Residual Learning for Image Recognition[J]. IEEE, 2016.
6. Wavelet Transform Time-Frequency Image and Convolutional Network-Based Motor Imagery EEG Classification[J]. IEEE Access, 2019, 7:6084–6093.
7. Bae W, Yoo J, Ye J C. Beyond Deep Residual Learning for Image Restoration: Persistent Homology-Guided Manifold Simplification[J]. IEEE, 2017.
8. He K, Zhang X, Ren S, et al. Delving Deep into Rectifiers: Surpassing Human-Level Performance on ImageNet Classification[J]. IEEE Computer Society, 2015.
9. Howard A G, Zhu M, Chen B, et al. MobileNets: Efficient Convolutional Neural Networks for Mobile Vision Applications[J]. 2017.
10. Sandler M, Howard A, Zhu M, et al. MobileNetV2: Inverted Residuals and Linear Bottlenecks [C]// 2018 IEEE/CVF Conference on Computer Vision and Pattern Recognition (CVPR). IEEE, 2018.
11. Li X, Ding L, Li W, et al. FPGA accelerates deep residual learning for image recognition[C]// 2017 IEEE 2nd Information Technology, Networking, Electronic and Automation Control Conference (ITNEC). IEEE, 2017.
12. Galdran A, Chakor H, Alrushood A A, et al. Automatic classification and triage of diabetic retinopathy from retinal images based on a convolutional neural networks (CNN) method [J]. Acta Ophthalmologica, 2019, 97.
13. Han K, Wang Y, Tian Q, et al. GhostNet: More Features From Cheap Operations[C]// 2020 IEEE/CVF Conference on Computer Vision and Pattern Recognition (CVPR). IEEE, 2020.
14. Krizhevsky A, Sutskever I, Hinton G. ImageNet Classification with Deep Convolutional Neural Networks[J]. Advances in neural information processing systems, 2012, 25(2).
15. Guo X, Li S, Zhang J, et al. PFLD: A Practical Facial Landmark Detector[J]. 2019.
16. Gao P, Lu K, Xue J, et al. A Coarse-to-Fine Facial Landmark Detection Method Based on Self-attention Mechanism[J]. IEEE Transactions on Multimedia, 2020, PP(99):1–1.
17. Burgos-Artizzu X P, Perona P, P Dollár. Robust Face Landmark Estimation under Occlusion [C]// IEEE International Conference on Computer Vision. IEEE, 2014.
18. Deng Z, Li K, Zhao Q, et al. Effective face landmark localization via single deep network [J]. 2017.
19. Hang, Zhao, Orazio, et al. Loss Functions for Image Restoration With Neural Networks [J]. IEEE Transactions on Computational Imaging, 2017.

Nonlinear Filter Combined Regularization of Compressed Sensing for CT Image Reconstruction

Yang Ding, Zhirong Cui, Hanxiu Dai, and Jian Dong

1 Introduction

Computed tomography (CT) has been widely used as one of the leading imaging modalities in medicine, biology, and industry. Cross-sectional images/volumes are obtained through reconstruction from projection data. X-ray photons radiate the object and the projection data is computed by measuring the non-attenuated portion which reaches detectors located on the opposite side of the X-ray source. The reconstructed CT image represents the attenuation distribution of X-ray photons [1–4]. The wide usage of CT undeniably improved the efficiency of disease diagnosis. However, recent discoveries regarding the potentially harmful effects of X-ray radiation including genetic and cancerous diseases have raised growing concerns among patients and medical physics community [5, 6]. A trend of leading the research in a direction of low radiation in medical diagnosis devices is increasingly expected. Three new concepts of CT device have been proposed and discussed, which are sparse-view CT, interior CT, and low-dose CT [7–13]. Three strategies

Y. Ding
Engineering Training Center, Tianjin University of Technology and Education, Tianjin, China

Z. Cui
Tianjin Nanyang Industrial School, Tianjin, China

H. Dai
College of Automation and Electrical Engineering, Tianjin University of Technology and Education, Tianjin, China

J. Dong (✉)
Tianjin Key Laboratory of Information Sensing and Intelligent Control, Tianjin University of Technology and Education, Tianjin, China

L. Meng (ed.), *International Conference on Cloud Computing and Computer Network*, Signals and Communication Technology,
https://doi.org/10.1007/978-3-031-47100-1_4

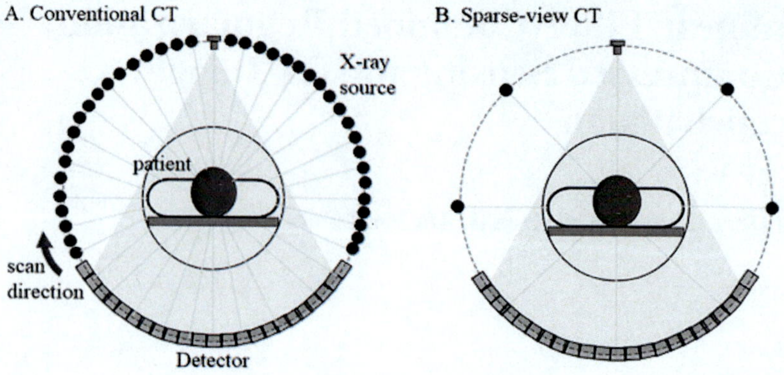

Fig. 1 Two CT imaging configurations of the conventional full scan CT and sparse view CT

above can all reduce patient dose significantly. In this study, we mainly elaborate on our novel ideas in image reconstruction of sparse view CT. Figure 1 shows two CT imaging configurations of the conventional full scan CT and sparse view CT.

The sparse-view CT refers to CT in which the measured number of projection data is reduced to less than 100 times determined by applications and required image quality. The reduction of projection data leads to benefits such as reducing patient dose, reducing scan time, and improving time resolution in cardiac CT. Since the sparse-view CT measures only incomplete projection data, classical image reconstruction methods such as filtered back projection (FBP) and algebraic reconstruction technique (ART) are not successful in producing clinically feasible images anymore. Because, in most cases, severe streak artifacts appear on the reconstructed image. The situation changed dramatically thanks to the discovery of compressed sensing (CS) as a new technique to solve inverse problems [9, 14, 15]. The goal of CS is to accurately reconstruct a signal or an image from insufficient measured data that is under-sampled below the Nyquist rate violating the conditions of Shannon sampling theorem.

As the framework of CS, the CT image reconstruction problem usually can be formulated as minimizing a cost function. The cost function typically consisted of two terms, the data-fidelity term and the penalty (regularization) term. Extensive experiments have shown that the design of regularization term in the cost function plays a critical role in reconstructing successful and accurate images [16–19].

The objective of our study was to apply nonlinear filter-based compressed sensing to sparse-view CT image reconstruction so as to obtain feasibly diagnosing image. In order to promote processing speed, we further developed row-action type algorithm thanks to proximal splitting theory. This chapter is organized as follows. In Sect. 2, topic context and literature review. In Sect. 3, the proposed nonlinear filter-based CS framework and associated iterative reconstruction methods are explained. In Sect. 4, practical CT image studies (dental, chest, and cranial CT image) performed to evaluate the proposed algorithm are shown. In Sect. 5, conclusion and discussion are described.

2 Related Works

In the literature, different approaches have been investigated to construct the regularization term of the cost function. One direction is to use a reference image for constructing the regularization term. The reference image could be a previously obtained data or other modalities data of the same patient, or the same body part image of other patients. Mameuda et al. proposed an approach for reconstructing PET/SPECT images with superior noise/contrast properties using template images estimated from CT/MRI [20]. Subsequently, Rashed and Kudo reduced the required priori information to only a set of intensity values and obtained the artifacts suppressed image from sparse projection data [21]. However, it is not always easy and practical to collect the reference image as a priori information and research of formulating the regularization term only utilizing information from the current scan attracted much interest. A series of quadratic-form ($\ell 2$ norm) regularization based on the Markov random field (MRF) model which describes the statistical distribution of a pixel given its neighbors have been proposed for iterative image reconstruction [16, 22–24]. The above works of literature all showed the ability to improve reconstructed image quality in terms of both noise suppression and edge preservation. However, limitations of the quadratic-form regularization had been subsequently recognized that it fails to take the discontinuities of images into account, thus it may lead to over smoothed of edges or fine structures in the reconstructed images. To overcome the limitation, Li et al. proposed an accurate iterative reconstruction algorithm based on the non-quadratic ℓp norm regularization term where $p = 1.1$. The optimum image was obtained by solving the duality problem of nonlinear programming, and noticeable gains over the traditional ART method were observed [25]. Several other literature also investigated the non-quadratic functions for noise suppressing and edge-preserving [26–29]. The above literature proved that the ℓp norm where $p < 2$ has a superior ability in image reconstruction from ill-posed inverse problems.

Total variation (TV) minimization has become the most investigated method in CS area of image reconstruction. The cost function of TV can be interpreted as $\ell 1$ norm of the magnitude image of intensity gradient. TV minimization was introduced for the first time as an efficient image-denoising method by Rudin, Osher, and Fatemi, and it is also called "ROF model" [30]. Subsequently, Sidky et al. applied the TV minimization to a variety of image reconstruction problems with incomplete projection data [31, 32]. They also developed a number of different iterative methods for the TV minimization, in which the overall structure is to repeat one iteration of the ART method and update the image towards the descent direction of the TV alternately. There are also other works on the use of TV [33, 34]. It has been proved that the TV minimization methods perform very well with simulated data, where pixel intensity values within the same region are almost uniform. However, this is not always true when it goes to real data. The images reconstructed by TV minimization method are known to suffer from some distortion, such as artificial patches on the image, improper serrate edges on object parts, and texture missing on smooth parts, which reduces the probability of application on practical CT image reconstruction system [17, 35].

In this chapter, we proposed a cost function in the form of subtraction operation between solving data and nonlinear filter processed data, which is followed by a data fidelity constraint term. Namely, we put our focus on nonlinear image filters instead of linear filters like TV for constructing regularization terms. Nonlinear filters, median filter, bilateral filter, and nonlocal weighted means (NL wmeans) filter were tested. Compared to median filter and bilateral filter, NL wmeans filter considers not only distance factor but also pixel intensity similarity with central pixel (or central area), so it is expected to present better reconstructions. NL wmeans filter was introduced by Buades et al. for image denoising [36, 37]. Inspired by its success in image processing scenarios, researchers further extended it to investigate the NLM-based regularizations for several inverse problems [38–40], but most of them utilized other scan images or neighboring frame images. Zhang et al. explored incorporating an NLM-based generic regularization for iterative reconstruction for low-dose CT, which in the regularization only utilizes information of the current scan [41]. They evaluated the NL wmeans-based regularization using both digital and physical phantom datasets, and observed noticeable gains. However, they only explored the MRF model-based regularization term in quadratic potential function. In this work, the proposed $\ell 1$ norm regularization can be expected to further improve the performance. Furthermore, we also proposed the mixed regularization (NL wmeans based $\ell 1$ norm item and TV item) in order to avoid the isolated points of error. As same as common statistical iterative image reconstruction methods, the proposed nonlinear filter-based CS method also takes an enormous time until the reconstructed image converges. In order to promote program execution efficiency, we split the cost function to block functions and employed proximal splitting theory. The row-action type image reconstruction algorithm was constructed.

3 Methodology

3.1 Problem Definition

The aim of CT imaging is to reconstruct an object from measured projection data. When iterative techniques are used for image reconstruction, this problem can be formulated as solving the linear equation. $A\vec{x} = \vec{b}$ where $\vec{x} = (x_1, x_2, \ldots, x_J)^T$ is the discrete representation of the object, $A = \{a_{ij}\}$ $(i = 1, 2, \ldots, I, j = 1, 2, \ldots, J)$ is the $I \times J$ system matrix and $\vec{b} = (b_1, b_2, \ldots, b_I)^T$ is the measured projection data. When the dimension of $\vec{b}$ is smaller than the dimension of $\vec{x}$ (i.e. $I < J$), this problem becomes underdetermined (sparse view CT reconstruction problem) and to obtain an accurate image reconstruction becomes challenging.

To overcome this problem, we proposed the following regularization term constrained by data fidelity term.

$$\text{minimize}\, f\left(\vec{x}\right) = \sum_{j=1}^{J} \left|x_j - M_j\left(\vec{x}\right)\right|_P^P \text{subject to}\, A\vec{x} = \vec{b} \tag{1}$$

where $M\left(\vec{x}\right) = \left(M_1\left(\vec{x}\right), M_2\left(\vec{x}\right), \ldots, M_j\left(\vec{x}\right)\right)$ is the output image after applying nonlinear filter on image $\vec{x}$.

The single ℓp norm regularization remains some isolated points of error in some cases, we proposed a more comprehensive TV-added regularization called mixed regularization.

$$\text{minimize}\, f\left(\vec{x}\right) = \beta \sum_{j=1}^{J} \left|x_j - M_j\left(\vec{x}\right)\right|_P^P + \gamma \left\|x_j - M_j\left(\vec{x}\right)\right\|_{TV} \text{ subject to}\, A\vec{x} = \vec{b} \tag{2}$$

where β and γ are hyper-parameters used to control the strength of penalty function part. And TV term is defined below, where (m, n) denotes the concerned pixel.

$$\|\vec{x}\|_{TV} = \sum_m \sum_n \sqrt{(x_{m,n} - x_{m-1,n})^2 + (x_{m,n} - x_{m,n-1})^2} \tag{3}$$

The parameter p in Eq. (1) is the order of norm. In classical solution techniques of inverse problems, $p = 2$ (called $\ell 2$ norm) is normally used. However, in some literature, $1 < p < 2$ has been proved to be more rigorous and effective in picking up sparse solutions than $\ell 2$ norm [20, 25]. In order to evaluate the power in picking up sparse solutions, an intuitive example of an image $\vec{x} = (x_1, x_2)^T$ with only two-components and projection data $\vec{b} = (b_1)$ with a single-component, is presented here. Figure 2 illustrates equi-contour lines corresponding to the ℓp norm with $p = 0$, $p = 1$, and $p = 2$. The brunet line $\mathcal{L}$ denotes the linear equation $a_{11}x_1 + a_{12}x_2 = b_1$. In

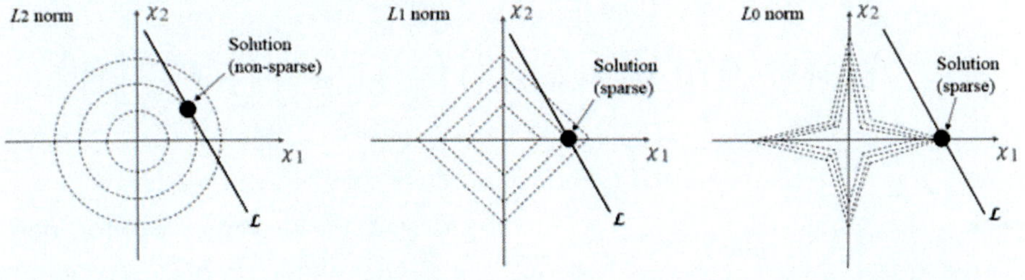

Fig. 2 Equi-contour lines corresponding to $\ell 2$, $\ell 1$, $\ell 0$ norm minimization in the case where $\vec{x} = (x_1, x_2)^T$ is a 2D vector

this case, the intersection point between $\mathcal{L}$ and x_1 (or x_2) axis is the sparse solution that we would like to reconstruct. From the equi-contour lines shown in Fig. 2, it is clear that the intersection point approaches coordinate axes (sparse solutions) as the cost reduces in $\ell 0$ and $\ell 1$ norm mode. Whereas the intersection point converges far from coordinate axes as cost reduces in $\ell 2$ norm mode. Accordingly, the $\ell 0$ norm and $\ell 1$ norm are naturally expected to possess a stronger power in picking up sparse solutions than the $\ell 2$ norm, and furtherly in higher-dimension problems, the difference between $\ell 0/\ell 1$ and $\ell 2$ norms could be significant. In this study, we performed $\ell 0$, $\ell 1$ and $\ell 2$ norms of proposed method and confirmed the reconstructed images.

The constrained minimization problem in Eq. (1) can be formulated equally to the unconstrained minimization as follows. The cost function then consists of data fidelity term and regularization term (penalty term).

$$\text{minimize} f\left(\vec{x}\right) = \left\|A\vec{x} - \vec{b}\right\|^2 + \beta \sum_{j=1}^{J} \left|x_j - M_j\left(\vec{x}\right)\right|_P^P \tag{4}$$

where β is the hyper-parameter used to control the strength of penalty function part.

3.2 *Iterative Reconstruction Methods*

In this section, we will discuss the iterative reconstruction algorithm for cost function (4). The main obstacle in using the $\ell 0/\ell 1$ norm is the lack of convexity and differentiability of the regularization term. Overcoming this drawback was made possible by splitting the object function to block functions. Proximal splitting theory was used sequentially in image restorations between block functions. The cost function was split as below.

$$f\left(\vec{x}\right) = \left\|A\vec{x} - \vec{b}\right\|^2 + \beta \sum_{j=1}^{J} \left|x_j - M_j\left(\vec{x}\right)\right|_P^P = f_1\left(\vec{x}\right) + \beta f_2\left(\vec{x}\right) \tag{5}$$

First, proximal operator was applied to $f_1\left(\vec{x}\right)$.

$$\vec{a}\left(\vec{x}^{(k)}\right) = \text{prox}_{\alpha f_1}\left(\vec{x}^{(k)}\right) = \arg\min_{\vec{x}} \left(\left\|A\vec{x} - \vec{b}\right\|^2 + \frac{1}{2\alpha}\left\|\vec{x} - \vec{x}^{(k)}\right\|^2\right) \tag{6}$$

where $\vec{a}\left(\vec{x}^{(k)}\right)$ can be treated as the intermediate solution, and α is the step size which takes a small enough value in order to guarantee the convergence of function. The result of $\vec{a}\left(\vec{x}^{(k)}\right)$ below is derived by $\partial \text{prox}_{\alpha f_1}\left(\vec{x}^{(k)}\right)/\partial \vec{x} = 0$, which represents exactly the same form with the conventional steepest descent method.

$$\vec{a}\left(\vec{x}^{(k)}\right) = \vec{x}^{(k)} - 2\alpha A^T\left(A\vec{x}^{(k)} - \vec{b}\right) \tag{7}$$

Next, proximal operator was applied to $f_2\left(\vec{x}\right)$ using the intermediate solution $\vec{a}\left(\vec{x}^{(k)}\right)$.

$$\vec{x}^{(k+1)} = \text{prox}_{\alpha f_2}\left(\vec{a}\left(\vec{x}^{(k)}\right)\right) = \arg\min_{\vec{x}}\left(\beta\left|\vec{x} - M\left(\vec{x}\right)\right|_p^p + \frac{1}{2\alpha}\left\|\vec{x} - \vec{a}\left(\vec{x}^{(k)}\right)\right\|^2\right) \tag{8}$$

However, $M\left(\vec{x}\right)$ is an unknown image vector since it is related to the unknown $\vec{x}$. When step size α takes a value small enough, $\vec{x}$ approximates infinitely to $\vec{a}\left(\vec{x}^{(k)}\right)$, so the constant approximation was used as $M\left(\vec{a}\left(\vec{x}^{(k)}\right)\right)$. Next, it is necessary to discuss the situation because of different p value.

In $\ell 2$ norm mode, proximal operation formula is differentiable. By setting $\partial \text{prox}_{\alpha f_2}\left(\vec{a}\left(\vec{x}^{(k)}\right)\right)/\partial\vec{x} = 0$, the solution could be calculated easily.

$$\vec{x}^{(k+1)} = \frac{\vec{a}\left(\vec{x}^{(k)}\right) + 2\alpha\beta M\left(\vec{a}\left(\vec{x}^{(k)}\right)\right)}{1 + 2\alpha\beta} \tag{9}$$

In $\ell 1$ norm mode, proximal operation formula is not differentiable, and it becomes a little complex to acquire solution. But a closed form of solving this kind of problem called "soft thresholding" can be applied. The result is expressed as below, where j denotes pixel number.

$$x_j^{(k+1)} = \begin{cases} a_j\left(\vec{x}^{(k)}\right) - \alpha\beta & \left(a_j\left(\vec{x}^{(k)}\right) - M_j\left(\vec{a}\left(\vec{x}^{(k)}\right)\right) > \alpha\beta\right) \\ a_j\left(\vec{x}^{(k)}\right) + \alpha\beta & \left(a_j\left(\vec{x}^{(k)}\right) - M_j\left(\vec{a}\left(\vec{x}^{(k)}\right)\right) < -\alpha\beta\right) \\ M_j\left(\vec{a}\left(\vec{x}^{(k)}\right)\right) & (\text{otherwise}) \end{cases} \tag{10}$$

The same as $\ell 1$ norm mode, $\ell 0$ norm mode is not differentiable either. So we used another closed form called "hard thresholding" to obtain the updating formula. The result is expressed below.

$$x_j^{(k+1)} = \begin{cases} a_j\left(\vec{x}^{(k)}\right) & \left(a_j\left(\vec{x}^{(k)}\right) - M_j\left(\vec{a}\left(\vec{x}^{(k)}\right)\right) < -\sqrt{2\alpha\beta}\right. \\ & \left. or\ a_j\left(\vec{x}^{(k)}\right) - M_j\left(\vec{a}\left(\vec{x}^{(k)}\right)\right) > \sqrt{2\alpha\beta}\right) \\ M_j\left(\vec{a}\left(\vec{x}^{(k)}\right)\right) & (\text{otherwise}) \end{cases} \tag{11}$$

Accordingly, the algorithm could be summarized as Table 1.

Table 1 The flow of the proposed algorithm

Step	Procedure
Step 1	(Initialization) set the initial image $\vec{\mathbf{x}}^{(0)}$ and set the iteration number as k←**0**.
Step 2	(Computation of intermediate image) compute the intermediate image $\vec{a}\left(\vec{x}^{(k)}\right)$ from the current image $\vec{x}^{(k)}$ by Eq. (7).
Step 3	(Image update) calculate the updated image $\vec{x}^{(k+1)}$ from the intermediate image $\vec{a}\left(\vec{x}^{(k)}\right)$. The calculation process is shown in Eqs. (9), (10), and (11) respectively based on norm mode ℓ2, ℓ1 and ℓ0.
Step 4	(Convergence check) increase the iteration number as k + 1← ***k***, and go to [Step 2] until reaching to a stopping criteria.

3.3 Algorithm Acceleration: Row-Action Structure

In the proposed algorithm in Sect. 3.2, the image update was achieved based on Eq. (7), and for one update, a whole projection data was utilized. So the image restoration is one of the simultaneous-iterative-type method, which has a poor convergence rate. In order to achieve algorithm acceleration, we further split the block function $f_1\left(\vec{x}\right)$ of Eq. (5) to row-action type.

$$f_1\left(\vec{x}\right)=\sum_{i=1}^{I} f_i\left(\vec{x}\right), f_i\left(\vec{x}\right)=\left(\vec{a}_i^T\vec{x}-b_i\right)^2 \tag{12}$$

Similarly, proximal operator was applied to Eq. (12).

$$\vec{x}^{(k,\ i+1)}=\text{prox}_{\alpha f_i}\left(\vec{x}^{(k,i)}\right)=\arg\min_{\vec{x}}\left(\left\|\vec{a}_i^T\vec{x}-b_i\right\|^2+\frac{1}{2\alpha}\left\|\vec{x}-\vec{x}^{(k,i)}\right\|^2\right) \tag{13}$$

And it could be expressed as the following constrained minimization.

$$\underset{\vec{x}}{\text{minimize}}\, f_i\left(\vec{x},z_i\right)=\frac{1}{2\alpha}\left\|\vec{x}-\vec{x}^{(k,i)}\right\|^2+(z_i-b_i)^2 \text{ subject to } z_i=\vec{a}_i^T\vec{x},\ \vec{x}\geq 0, \tag{14}$$

Then a Lagrange function is constructed as below where λ_i is the Lagrange coefficient.

$$L_i\left(\vec{x},z_i,\lambda_i\right)=\frac{1}{2\alpha}\left\|\vec{x}-\vec{x}^{(k,i)}\right\|^2+(z_i-b_i)^2+\lambda_i\left(z_i-\vec{a}_i^T\vec{x}\right) \tag{15}$$

Take a derivative based on variables $\vec{x}, z_i, \lambda_i$ respectively, and the iterative image update equation can be obtained below.

$$\vec{x}^{(k,i+1)} = \vec{x}^{(k,i)} + \alpha\lambda_i \vec{a}_i; \lambda_i = \frac{b_i - \vec{a}_i^T \vec{x}^{(k,i)}}{1/2 + \alpha\|\vec{a}_i\|^2} \tag{16}$$

In Sect. 3.2, the image update was achieved according to two separate steps. First, image was reconstructed using the steepest descent method based on the data fidelity term. Second, the $\ell 0/\ell 1/\ \ell 2$ norm was minimized using thresholding methods or a fixed equation. These two steps were repeated in an alternating manner. In the algorithm acceleration part (row-action structure), the same framework was utilized. The only difference was that image reconstruction and norm minimization were not performed in an alternating manner, but to insert norm minimization procedure empirically in every appropriate times of image reconstruction.

4 Experimental Results

In this study, sparse-view medical CT image reconstruction was implemented. It was applied on practical dental CT, chest CT, and cranial CT images. The proposed nonlinear filter-based CS algorithm was tested. Based on the working mechanism of different nonlinear filters, median filter, bilateral filter and nonlocal weighted means filter were combined in regularization term for comparison [42–44]. In order to avoid the isolated points of errors in reconstructed images, we also tested TV norm mixed regularization. We further accelerated implementation efficiency by developing row-action type image update algorithm based on the proximal splitting theory. Projection data was computed by parallel-beam geometry over 180°. The processing was executed based on C-language program. The performance of the computer that we used was as follows: an Intel(R) Core(TM) i7-4770 central processing unit, running at 3.40 and 3.40 GHz and a Windows 8.1 Pro operating system.

The pixel size of dental images shown in Figs. 3 and 4 is 512 × 512 and the 48 views projection data was used for the dental image reconstruction. The result images of Fig. 3 were shown for effect comparison of $\ell 2$, $\ell 1$, and $\ell 0$ norm. For a fair comparison, we utilized the relatively simpler median filter combined regularization. Median filter window size was all set in 7 × 7, and 1000 iterations were implemented in each case. The hyper-parameter β was set to play the same smoothing effect in each case. The images were reconstructed by Fig. 3a median filter-based $\ell 2$ norm regularization, Fig. 3b median filter-based $\ell 1$ norm regularization, and Fig. 3c median filter-based $\ell 0$ norm regularization. Image Fig. 3a showed blurry edges and teeth structure were also over smoothed that the vast majority of dentin part was lost. Image Fig. 3b presented smooth object with clear edges and intact texture. Image Fig. 3c showed clear edges but isolated points of error were also obvious. The results of Fig. 3 showed that $\ell 1$ norm regularization was superior for smoothing

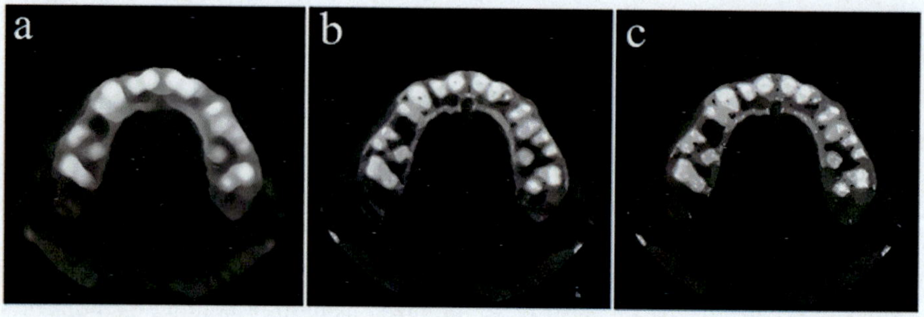

Fig. 3 Result images of (**a**) median filter based ℓ2 norm regularization, (**b**) median filter based ℓ1 norm regularization, and (**c**) median filter based ℓ0 norm regularization. 1000 iterations were implemented in each case

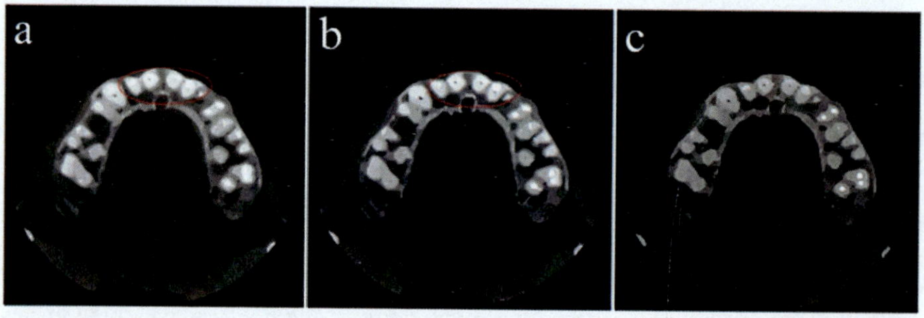

Fig. 4 Result images of (**a**) median filter based ℓ1 norm regularization, (**b**) NL wmeans filter based ℓ1 norm regularization, and (**c**) bilateral filter based ℓ1 norm regularization. 1000 iterations were implemented in each case

object, extracting object edge and enhancing image texture when used in sparse-view CT reconstruction. So we concentrated to discuss ℓ1 norm regularization effect of different nonlocal filters in Fig. 4.

Images of Fig. 4 shows a. median filter-based ℓ1 norm regularization, b. NL wmeans filter-based ℓ1 norm regularization, and c. bilateral filter-based ℓ1 norm regularization. 1000 iterations were implemented in each case. The result images of Fig. 4 demonstrated that combining different nonlocal filters in regularization term lead to different reconstruction images. Figure 4a is the same image with Fig. 3b which presented clear edges and some isolated points of error. Figure 4b showed much clearer edges of teeth structure especially circled part with red line. Isolated points of error were also removed and image texture was well-preserved. Figure 4c showed clear edges of teeth structure, but the image texture was over smoothed that pixel intensity was degraded to just like a certain value. Of the three kinds of combined nonlocal filters, NL wmeans filter showed the best results.

Because dental CT image presents relatively simple structural features, the proposed method may work easily. Therefore, we further implemented the NL wmeans filter combined ℓ1 norm regularization to chest CT image. The pixel size

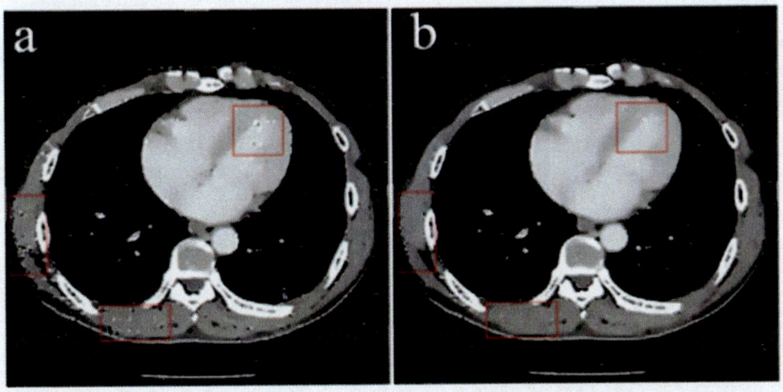

Fig. 5 (**a**) Reconstructed chest image under the implementation of NL wmeans filter combined ℓ1 regularization. (**b**) Reconstructed image of the mixed regularization (NL wmeans filter combined ℓ1 norm regularization and TV norm regularization)

of chest image in Fig. 5 is 512 × 512, and 72 projection views were acquired and used for the reconstruction. Figure 5a is a reconstructed chest image under the implementation of NL wmeans filter combined ℓ1 regularization. As a result, isolated points of error occurred among the resulting image. So we implemented the mixed regularization whose cost function is shown in formula (2). Besides the NL wmeans filter combined ℓ1 norm regularization term, TV norm was also considered in iterative reconstruction. Figure 5b shows the result image of the mixed regularization. Isolated points of error were obviously removed. So the mixed regularization showed effectiveness in reconstructing the more complex images.

Algorithm acceleration was carried out by splitting the image reconstruction procedure to row-action type. In non-row-action type reconstruction, only one image restoration was carried out in one iteration; however, projection view number times of image restorations could be achieved in one iteration of the row-action type reconstruction. So the image convergence speed could be improved greatly. In Fig. 6, the acceleration effectiveness was investigated using cranial CT image. The pixel size of cranial CT image shown in Fig. 6 is 512 × 512 and 72 views projection data was used for reconstruction. The algorithm we used here was NL wmeans filter-based CS (ℓ1 norm). The images in the top row of Fig. 6 were reconstructed by non-row-action type reconstruction, and they are the results of 100 iterations, 200 iterations, and 300 iterations, respectively. The image convergence was so slow that 5000 iterations were essential until the image converged. The total cost time of the 5000 iterations was 267 min. The images in the bottom row of Fig. 6 were reconstructed by the accelerated row-action type algorithm. They are the results of 1 iteration, 2 iterations, and 3 iterations. Only 5 iterations were needed until the image converged. The total cost time of 5 iterations was about 8 min. The computation time was reduced to 1/26 without influencing image quality. Root mean squared error (RMSE) was used to evaluate the image quality, and they were shown in the lower left corner of each image.

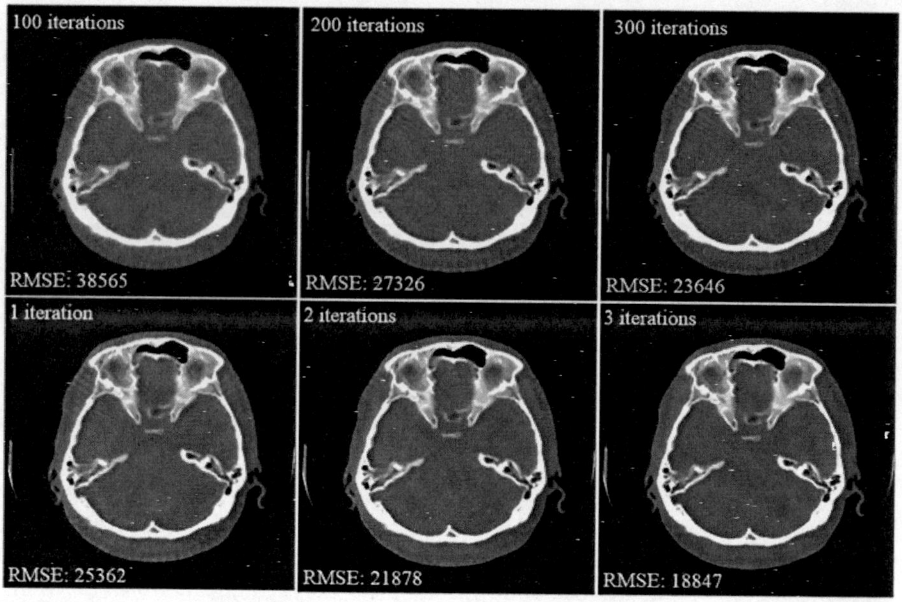

Fig. 6 The images in the top row were reconstructed by non-row-action type reconstruction, and they are the results of 100 iterations, 200 iterations, and 300 iterations, respectively. The images in the bottom row were reconstructed by the accelerated row-action type algorithm. They are the results of 1, 2 and 3 iterations

5 Conclusions and Discussions

Nonlinear filter-based compressed sensing was proposed and applied on sparse-view CT reconstruction in this chapter. Median filter, bilateral filter, and nonlocal weighted means filter were combined in regularization item as nonlinear filters. $\ell 0$, $\ell 1$, and $\ell 2$ norm of the proposed method were carried out, respectively, in this study. Nonlinear filter-based $\ell 1$ norm regularization added by TV norm term which was called mixed regularization was also proposed for reconstructing more complex images. The proposed method showed the superior effect in image smoothing, edge extracting and texture depicting. It was also proved that the $\ell 1$ norm usually works better than $\ell 0$ and $\ell 2$ norm, and nonlocal weighted means filter usually performs better than other two nonlinear filters.

The idea that combining the nonlinear filter into CS framework presented, respectively, high novelty. The main reason of not being considered before can be summarized as follows: (1) It was not expected to achieve much image quality improvement although the nonlinear filter be combined in. (2) Non-convexity and non-differentiability of object function evolved when using nonlinear filter, and function minimization became impossible. We solved the problem by proximal splitting theory and proved that the nonlinear filter-based compressed sensing was highly effective in sparse-view CT reconstruction. Furthermore, we carried out calculation acceleration by enhancing the reconstruction to row-action structure, and applied it on the practical medical CT images. This showed great value of our method.

We showed the application possibility of CS in commercial CT. Besides, the technique is being needed in many other imaging devices. In micro-CT used for small animal imaging and non-destructive testing, X-ray exposure time of each projection view is longer compared to that in commercial medical CT scanner. What's more, transferring the measured projection data to computer also takes a large amount of time. So reducing the projection views possesses a significant benefit. In cardiac CT, so-called cardiac gating is used to decrease motion artifacts. It is necessary to obtain the projection data over n heartbeats, but the number n becomes too large in the consideration of patient dose and no movement tolerance. So the image reconstruction must be performed from the sparse view of projection data [45, 46]. The technique is also expected to be used in flat panel detector CT and 3D angiography.

The nonlinear filters, including the bilateral filter and non-local weighted means filter we used in this study, usually have more than one parameter. Appropriate adjustment of the parameters is related to the image quality and this is a sticking point of this study. Consideration of reducing parameter number and automatically parameter setting becomes our future work. We are also attempting to apply the proposed method on denoising problems and image reconstructions from noisy projection data, what we call low-dose CT in this study.

Acknowledgments This work is supported by scientific research project of Tianjin Education Commission (Grants No. 2021KJ012) and Innovation & Entrepreneurship Project for College Students (Grants No.202210066008).

References

1. Takanori, M., Takeshi, N., Yoshinori, F.,et al. 2023. RADIATION DOSE REDUCTION AT LOW TUBE VOLTAGE WITH CORONARY ARTERY BYPASS GRAFT COMPUTED TOMOGRAPHY ANGIOGRAPHY BASED ON THE CONTRAST NOISE RATIO INDEX. Radiation Protection Dosimetry 6(6),(2023).
2. Mansouri, M., Choukri, A., Semghouli, S., Talbi, M., Eddaoui, K., Saga, Z.: Size-specific dose estimates for thoracic and abdominal computed tomography examinations at two moroccan hospitals. Journal of Digital Imaging 35(6), 1648–1653(2022).
3. Frandon, J., Akessoul, P., Hamard, A.,et al.: Comparison of acquisition and iterative reconstruction parameters in abdominal computed tomography-guided procedures: a phantom study. AME Publishing Company 2022(1).DOI:https://doi.org/10.21037/QIMS-21-328 (2022).
4. Herman, GT.: Image reconstruction from projections: implementation and applications. Springer, (1979).
5. Brenner, D. and Hall, J.: Computed Tomography – An increasing source of radiation exposure N.Engl.J.Med 357, 2277–84 (2007).
6. Hall, E.J., Brenner, D.j.: Cancer risks from diagnostic radiology. The British Journal of Radiology 81, 362–378 (2008).

7. Siltanen, S., Kolehmainen, V., Jarvenpaa, S. et al: Statistical inversion for medical X-ray tomography with few radiographs: I. general theory. Phys Med Biol 48: 1437–1463 (2003).
8. Herman, G.T., Davidi, R.: Image reconstruction from a small number of projections. Inverse Problems 24 Article ID 045011 (2008).
9. Candes, E.J., Romberg, J., Tao, T.: Robust uncertainty principles: exact signal reconstruction from highly incomplete frequency information. IEEE Trans Inf Theory 52: 489–509 (2006).
10. Pan, X., Zou, Y., Xia, D.: Image reconstruction in peripheral and central regions-of-interest and data redundancy. Med Phys 32: 673–684 (2005).
11. Defrise, M., Noo, F., Clackdoyle, R. et al.: Truncated Hilbert transform and image reconstruction from limited tomographic data. Inverse Problems 22: 1037–1053 (2006).
12. Kudo, H., Suzuki, T., Rashed, E.A.: Image reconstruction for sparse-view CT and interior CT: Introduction to compressed sensing and differentiated backprojection. Quant Imaging Med Surg 3: 147–161 (2013).
13. Rampinelli, C., Origgi, D., Bellomi, M.: Low-dose CT: technique, reading methods and image interpretation. Cancer Imaging;12,548–56 (2013).
14. Donoho, D.L.: Compressed sensing. IEEE Trans Inf Theory 52,1289–306 (2006).
15. Candes, E.J., Wakin, M.B.: An introduction to compressive sampling. IEEE Signal Processing Magazine; 25,21–30 (2008).
16. Ouyang, L., Solberg, T., Wang, J.: Effects of the penalty on the penalized weighted least-squares image reconstruction for low-dose CBCT. Phys Med Biol 56,5535–52 (2011).
17. Tang, J., Nett, B., Chen, G.: Performance comparison between total variation (TV)-based compressed sensing and statistical iterative reconstruction algorithms. Phys Med Biol 54,5781–804 (2009).
18. Wang, J., Li T., Xing, L.: Iterative image reconstruction for CTCT using edge-preserving prior. Med Phys 36, 252–60 (2009).
19. Theriault-Lauzier, P., Chen, G.: Characterization of statistical prior image constrained compressed sensing II: application to dose reduction. Med Phys 40(2), 021902 (2013).
20. Mameuda, Y., Kudo, H.: New anatomical-prior-based image reconstruction method for PET/SPECT. Conference Record of 2007 IEEE Nuclear Science Symposium and Medical Imaging Conference, Paper No. M23-2 (2007).
21. Rashed, E.A., Kudo, H.: Intensity-based Bayesian framework for image reconstruction from sparse projection data. Med Imag Tech 27, 243–251 (2009).
22. Hebert, T., Leahy, R.: A generalized EM algorithm for 3-D Bayesian reconstruction from Poisson data using Gibbs priors. IEEE Trans Med Imaging 8, 194–202 (1989).
23. Sauer, K., Bouman, C.: A local update strategy for iterative reconstruction from projections. IEEE Trans Signal Process 41, 534–48 (1993).
24. Wang, J., Li, T., Lu, H., Liang, Z.: Penalized weighted least-squares approach to sinogram noise reduction and image reconstruction for low-dose X-ray computed tomography. IEEE Trans Med Imaging 25, 1272–83 (2006).
25. Li, M., Yang, H., Kudo, H.: An accurate iterative reconstruction algorithms for sparse objects: application to 3D blood vessel reconstruction from a limited number of projections. Phys Med Biol 47, 2599–2609 (2002).
26. Green, P.J.: Bayesian reconstruction from emission tomography data using a modified EM algorithm. IEEE Trans Med Imaging 9, 84–93 (1990).
27. Lange, K.: Convergence of EM image reconstruction algorithms with Gibbs priors. IEEE Trans Med Imaging 9, 439–46 (1990).
28. Bouman, C., Sauer, K.: A generalized Gaussian image model for edge-preserving MAP estimation. IEEE Trans Image Process 2, 296–310 (1993).
29. Charbonnier, P., Aubert, G., Blanc-Feraud, L., Barlaud, M.: Two deterministic half-quadratic regularization algorithms for computed imaging. In Proc. 1st IEEE ICIP (1993).
30. Rudin, L.I., Osher, S., Fatemi, E.: Nonlinear total variation based noise removal algorithms. Physica D: Nonlinear Phenomena 60, 259–268 (1992).

31. Sidky, E.Y., Kao, C.M., Pan, X.: Accurate image reconstruction from few-views and limited-angle data in divergent-beam CT. J X-ray sci Tech 14, 119–39 (2006).
32. Sidky, E.Y., Pan, X.: Image reconstruction in circular cone-beam computed tomography by constrained, total-variation minimization. Phys Med Biol 53, 4777–807 (2008).
33. Song, J., Liu, Q.H., Johnson, G.A., et al.: Sparseness prior based iterative image reconstruction for retrospectively gated cardiac micro-CT. Med Phys 34, 4476–83 (2007).
34. Ritschl, L., Bergner, F., Fleischmann, C., et al.: Improved total variation-based CT image reconstruction applied to clinical data. Phys Med Biol 56, 1545–61 (2011).
35. Fahimian, B.P., Mao, Y., Cloetens, P., Miao, J.: Low-dose x-ray phase-contrast and absorption ct using equally sloped tomography. Physics in Medicine and Biology 55, 5383(2010).
36. Buades, A., Coll, B., Morel, J.: A non-local algorithm for image denoising. IEEE Comput Vis Pattern Recognit 2, 60-5 (2005).
37. Buades, A., Coll, B., Morel J.: A review of image denoising algorithms with a new one. Multiscale Model Simul 4(2), 490–530 (2005).
38. Lou, Y., Zhang, X., Osher, S., Bertozzi, A.: Image recovery via nonlocal operators. SIAM J Sci Comput 42(2), 185–97 (2010).
39. Tian, Z., Jia, X., Dong, B., Lou, Y., Jiang, S.: Low-dose 4D CT reconstruction via temporal nonlocal means. Med Phys 38, 1359–65 (2011).
40. Ma, J., Zhang, H., Gao, Y., Huang, J., Liang, Z., Feng, Q., Chen, W.: Iterative image reconstruction for cerebral perfusion CT using a pre-contrast scan induced edge-preserving prior. Phys Med Biol 57, 7519–42 (2012).
41. Zhang, H., Ma, J., Wang, J., Liu, Y., Lu, H., Liang, Z.: Statistical image reconstruction for low-dose CT using nonlocal means-based regularization. Comp Med Imag Graph 38, 423–435 (2014).
42. Clark, D., Johnson, G.A., Badea, C.T.: Denoising of 4D Cardiac Micro-CT Data Using Median-Centric Bilateral Filtration. Proc SPIE Int Soc Opt Eng 2012, 8314 (2012).
43. Zheng, Y., Fu, H., Au, O.K., Tai, C.L.: Bilateral normal filtering for mesh denoising. IEEE Trans Vis Comput Graph. 2011 Oct 17(10),1521–30 (2011).
44. Dehghannasiri, R., Shirani, S.: A novel de-interlacing method based on locally-adaptive nonlocal-means. Signals, Systems and Computers, 2012 46th Asilomar Conference on, On page(s), 1708–12 (2012).
45. Patel, T.R., Todd, V., Kramer, C.M.,et al.: Great Debate: Computed tomography coronary angiography should be the initial diagnostic test in suspected angina. European Heart Journal, DOI:https://doi.org/10.1093/eurheartj/ehac597 (2023).
46. Patel, V.I., Roy, S.K., Budoff, M.J.: Coronary computed tomography angiography (ccta) vs functional imaging in the evaluation of stable ischemic heart disease. The Journal of invasive cardiology 33(5), E349–E354 (2021).

approximately 28,000 printers worldwide. Instead of causing damage, they discovered a flaw that allowed them to print a page that said, "This printer has been hacked" [6].

Printers are common devices whose networked use is enormously unsecured, perhaps due to an enrooted assumption that their services are somewhat negligible and, as such, unworthy of protection. [3]. With this clear assertion, printers still exhibit security flaws and lag behind other IoT and electronic devices that are beginning to conform to cybersecurity and data privacy regulations. In 2020, 41% of employees worked exclusively from home, and 31% of at-home employees cited security and privacy as one of their major concerns [7]. Hewlett-Packard (HP), however, has released a critical patch to address two vulnerabilities in the printer's network that had compromised 150 multifunction printers (MFPs) and tricked users into visiting a malicious website [8]. Network vulnerability is a flaw or loopholes in organizational procedures, hardware, or software that, if exploited by an outside threat, could lead to a security breach, or threaten the normal function of a certain network [9, 10].

According to the HP Jet Advantage Security Manager Solution overview, over 66% of IT managers now think that their workplace printers are infected with malware, and nearly 75% of CIOs anticipate that during the next few years, these devices will be the main targets of data intrusion [11].

Today, every Internet of Things (IOT) device connected to the network such as network printers and digital copiers faces a critical and underappreciated threat from the malicious use of printing protocols. Attacking client endpoints is now more commonplace like attacking the server infrastructure. Furthermore, all the features of a PC or server are also present on the MFP and networked printers (client endpoints). Therefore, this leaves us with the question of why are these devices still being installed at clients' locations without any security safeguards in place? Depending on the printer device and set-up, data can be transferred via USB/parallel cable or over the network. This case study will explore various attacks and mitigation techniques on network printing and local printers. Some devices allow printing via open-source and most popular printing protocols are directly supported by network printers which are found vulnerable to cyberattacks. These direct attacks on network printing protocols are found to be inevitable. The structure of this study is presented in Table 1.

The overall objective of this study is to create awareness, explore, and enlist the mitigation techniques, perhaps a growing list of measures that a user can take to lessen the likelihood of a breach that starts with office printers and digital copiers. The researchers would like to answer the following questions of (1) how secure would it be to implement a software-based print management solution, that claims to give an organization a complete control and accounting to end users having multiple printers and complex printing structures? (2) What about security issues in using serverless printing or cloud-based printing environment? Through this study, we can be able to assess and identify security vulnerabilities in printers and digital copiers connected to networks or USB, assess the security posture of a company using a software-based print management solution, determine the overall awareness or

Vulnerabilities in Office Printers, Multifunction Printers (MFP), 3D Printers, and Digital Copiers: A Gateway to Breach Our Enterprise Network

Eric B. Blancaflor and Allen James Montoya

1 Introduction

1.1 *Background of the Study*

Internet of Things (IoT) is a reality today, but is expected to become increasingly prevalent over time, bringing about the Internet of Everything [1]. From smartphones to computers, to home thermostats, security cameras, network printers, and coffee makers, the Internet of Things (IoT) is made up of millions of "smart," connected devices. However, the Internet of Things has both benefits and drawbacks in terms of security features such as preventing hackers from accessing the devices. In addition to the privacy and security concerns that these security flaws raise, hackers can use these connected devices to form "zombies of armies" or botnets, which are networks of devices that are infected with malware without the users' knowledge [2]. Nevertheless, devices connecting to local networks have been associated with numerous scandals. In one of the incidents, over trusting printers resulted in printer attacks, in which attackers recruited printers to form a Distributed Denial of Service (DDoS) swarm, imposing a paper DoS state on the computers, and exposing personal information [3, 4]. Cyberattacks have increased exponentially since the pandemic began in early 2020. According to IDC, IT security spending will exceed $6 trillion in 2021, with 70% of breaches beginning at the endpoint. CSO, however, stated that attacks on IoT (Internet of Things) devices tripled in the first half of 2019 and will continue to rise in the coming years if IoT remains vulnerable [5]. In August 2020, the CyberNews ethical hacker group gained access to

E. B. Blancaflor (✉) · A. J. Montoya
Mapua University, Makati, Philippines
e-mail: ebblancaflor@mapua.edu.ph

L. Meng (ed.), *International Conference on Cloud Computing and Computer Network*, Signals and Communication Technology,
https://doi.org/10.1007/978-3-031-47100-1_5

approximately 28,000 printers worldwide. Instead of causing damage, they discovered a flaw that allowed them to print a page that said, "This printer has been hacked" [6].

Printers are common devices whose networked use is enormously unsecured, perhaps due to an enrooted assumption that their services are somewhat negligible and, as such, unworthy of protection. [3]. With this clear assertion, printers still exhibit security flaws and lag behind other IoT and electronic devices that are beginning to conform to cybersecurity and data privacy regulations. In 2020, 41% of employees worked exclusively from home, and 31% of at-home employees cited security and privacy as one of their major concerns [7]. Hewlett-Packard (HP), however, has released a critical patch to address two vulnerabilities in the printer's network that had compromised 150 multifunction printers (MFPs) and tricked users into visiting a malicious website [8]. Network vulnerability is a flaw or loopholes in organizational procedures, hardware, or software that, if exploited by an outside threat, could lead to a security breach, or threaten the normal function of a certain network [9, 10].

According to the HP Jet Advantage Security Manager Solution overview, over 66% of IT managers now think that their workplace printers are infected with malware, and nearly 75% of CIOs anticipate that during the next few years, these devices will be the main targets of data intrusion [11].

Today, every Internet of Things (IOT) device connected to the network such as network printers and digital copiers faces a critical and underappreciated threat from the malicious use of printing protocols. Attacking client endpoints is now more commonplace like attacking the server infrastructure. Furthermore, all the features of a PC or server are also present on the MFP and networked printers (client endpoints). Therefore, this leaves us with the question of why are these devices still being installed at clients' locations without any security safeguards in place? Depending on the printer device and set-up, data can be transferred via USB/parallel cable or over the network. This case study will explore various attacks and mitigation techniques on network printing and local printers. Some devices allow printing via open-source and most popular printing protocols are directly supported by network printers which are found vulnerable to cyberattacks. These direct attacks on network printing protocols are found to be inevitable. The structure of this study is presented in Table 1.

The overall objective of this study is to create awareness, explore, and enlist the mitigation techniques, perhaps a growing list of measures that a user can take to lessen the likelihood of a breach that starts with office printers and digital copiers. The researchers would like to answer the following questions of (1) how secure would it be to implement a software-based print management solution, that claims to give an organization a complete control and accounting to end users having multiple printers and complex printing structures? (2) What about security issues in using serverless printing or cloud-based printing environment? Through this study, we can be able to assess and identify security vulnerabilities in printers and digital copiers connected to networks or USB, assess the security posture of a company using a software-based print management solution, determine the overall awareness or

Table 1 Conceptual framework

Input	Process	Output
The way printers and digital copiers are used and treated as common devices in an organization. The attack method an attacker may use against printers and digital copiers An attacker's capabilities before, during, and post-attack Recorded incidents with proof of principal attacks in the past	Survey questionnaire Interview vendors and service providers regarding security features of their products Conduct one of the attack simulations Conduct a reconnaissance using Shodan website Gather mitigation techniques and digital copier security attacks and exploits	Risks on businesses if printers are kept unsecured from social engineering attacks, security threat, and risk mitigation are not put in place. List of recommended countermeasures or proactive steps to prevent a print-based breach from occurring within the organization

understanding of IT users in securing network printers from various attacks, determine how unsecured print management has resulted in cyberattacks for various businesses or organizations in the past. Moreover, this study has surveyed IT managers, IT engineers, IT staff, and technical support vendors that manage companies with 50–1000 employees. Respondents came from the UAE, Saudi Arabia and the Philippines. To explore and understand the capabilities and security preparedness of their products against cyberattacks, we also included a few Middle Eastern manufacturers and service providers in the online survey.

2 Literature Review

2.1 You Over Trust Your Printer [3]

In this study, three attacks known as "Printjack" were put together by a group of Italian researchers to alert users of the serious repercussions of placing too much reliance in their printers. The researcher assessed a few potential effects of using raw printing using TCP port 9100. A network protocol that is regarded as the simplest and fastest and generally the most dependable network protocol used for printers. When they used Shodan, a search engine for interconnected devices, they were able to reveal the unethical practice of disclosing thousands of public IP addresses over the Internet that responded to the 9100-port query per country. The top 10 countries that responded to the port 9100 query were ranked by gross domestic profit (GDP), and Germany appeared to have the highest GDP and number of devices exposed. Publicly available results can provide a potential vulnerability that can be utilized to carry out the "Printjack" family of attacks. This may lead to Distributed Denial of Service (DDoS), paper Denial of Service (DoS), and privacy infringements. Each "Printjack" attack has been assessed using a qualitative risk assessment method based on ISO/IEC 27005:2018. By developing structured arguments and conducting technical experiments, the risk level of each attack has been calculated based on its likelihood and impact. Based on the results of the analysis, it was determined that all "Printjack" attacks are HIGH-RISK. Common vulnerabilities and exposures (CVEs) on printers from CVE Database were obtained by searching the word "printer" in the CVE website. Majority of these vulnerabilities allow remote execution of arbitrary code or commands. One example of vulnerability is CVE-2014-3741. Another supporting evidence that the case study had presented is the "Printjack" attack causing paper Denial of Service (DoS). Another "Printjack" attack scenario was presented as part of a social engineering strategy that addressed an inside threat. Typically, an end user sends a print job across a network and another person executes a Man In the Middle Attack (MITM) and eavesdrops on the printed material, which is a clear infringement of the user's privacy. A printer's network could theoretically be exploited by an attacker to retrieve data in plaintext, since printing data is not encrypted. To demonstrate the attack, the researchers placed an intermediary between the printer and the sender using Ettercap, and then Wireshark snooped on a PDF file that was being transmitted for printing. We learn from this study that printer vendors need to upgrade both their hardware and software to improve security and data handling. Furthermore, users and businesses ought to stop treating printers as an insignificant part of their daily computing, believing that printers cannot pose a real risk to data or themselves [3].

2.2 SoK: Exploiting Network Printers [12]

In this study, the researchers analyzed printer attacks on a large scale and developed a general methodology for analyzing printer security. Their methodology for assessing printer security led them to implement a tool called the Printer Exploitation Toolkit (PRET) to evaluate 20 printer models from different vendors and found that all of them are vulnerable to at least one of the tested attacks. Examples of these attacks include simple DoS attacks or sophisticated attacks aimed at extracting system files and printing jobs. Together with this analysis, they have revealed insights that enable cross-site printing using COR spoofing and advanced printing techniques. The study had explored the underlying threats on 3D printers. There has been a great deal of interest in 3D printers in recent years. Thus, 3D printers can also serve as attack targets or attack gadgets. Several researchers have shown how to modify design files (.STL files) to add voids and cavities that can damage the resulting object. Finally, they demonstrate how they can attack systems other than printers, such as Google Cloud Print or websites that process documents. On the other hand, the risks of malicious firmware updates are well understood and have been discussed by and in comparison, to other networked devices. Through simulated attacks, the researcher demonstrated how an attacker could run simple DoS attacks, access printing jobs, and even hack into company networks. There is a lack of security analysis tools used by printer manufacturers, and they do not take security incidents seriously [12].

2.3 *Printer Security Vulnerabilities and What You Can Do About It! [6]*

In this featured article, the author outlined the primary ways to secure printers and the exponential growth of cyberattacks since the pandemic started in early 2020. The IT Security expenditure in 2021 was estimated to exceed $6 Trillion, and IDC predicted 70% of breaches had begun at the endpoint devices. The term "endpoint" refers to any device connected to a network, which can include printers, PCs, thermostats, Ring doorbells, or security cameras. Printers, which are frequently disregarded by businesses, are among the largest network unknowns. Unless printers are properly configured with security features upon deployment, they can be used by cybercriminals as a backdoor into your network. After a printer is installed, it is rarely monitored unless it needs maintenance or consumable replacements. The author raised a vital point about why businesses don't take printer security seriously. The straightforward response is the lack of understanding on how vulnerable printers are from cyberattacks, or the complacency that a printer is protected by a corporate firewall, therefore it is impervious to external attacks. However, the fact is that firewalls are only one layer of security based on predetermined rules and are inadequate to avert attackers from accessing printers. The author had recommended using firmware assessment tools to verify the most recent version to assess the security health of printers and make modifications in real time to make sure the fleet of printers is running the most recent firmware [6].

3 Methodology

3.1 *Research Design*

This study focuses on gaining a deeper understanding regarding the awareness of IT users about the various social engineering attacks and security threats that can exploit printers in every business. Using frequency, magnitude, numbers, and graphs to test and confirm theories and assumptions regarding IT user understanding of unsecured printer devices and how attackers can move laterally across networks to access sensitive data by leveraging the inherent trust of compromised endpoints such as printers. Furthermore, a qualitative approach was used with the same online survey and the same respondents in an online mode of data collection.

The respondents to the study shall include IT and network management professionals. To collect relevant responses, the researchers used selective sampling. Our methodology entailed selecting IT users with a minimum of three (3) years of IT experience who manage networks regardless of gender and type of business. As the study focuses on a wide variety of printers, researchers selected IT professionals able to troubleshoot and monitor printer uptime and downtime. Whenever the machine breaks down or the print management software solution fails, the responders will coordinate with the service providers.

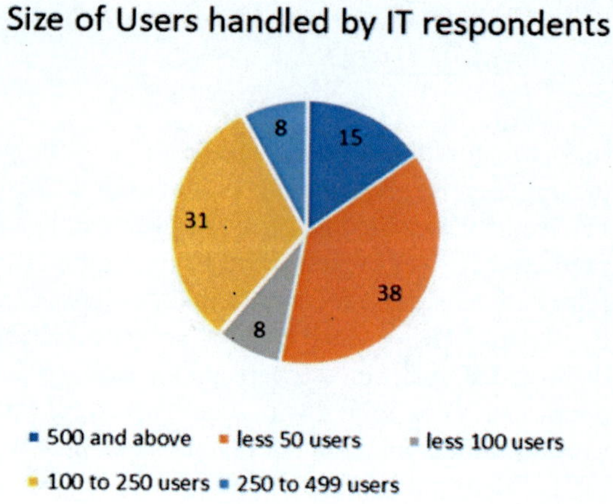

Fig. 1 Size of user organization

3.2 Results and Key Findings

This section summarizes the primary data collected through an online survey. Through brief explanations and graphs, the quantitative and qualitative data collected through the survey questionnaire are presented.

The first item in the online survey is to determine the years of experience of respondents. Regardless of gender or type of business, at least three (3) years of IT experience managing networks is required. It is critical to determine their exposure to network management and understanding of the importance of endpoint security in their organization. Thirteen (13) respondents have three (03) to thirty (30) years of IT experience. Figure 1 presents the network sizes of IT respondents ranging from less than 50 to 500 or more users. Less than 50 users account for 38% of IT respondents, with 100–250 users accounting for 31%. We also have users who manage networks of 250 or more. In this study, we looked at a small, medium, and large enterprise network that has office printers installed. The types of printers used in the IT respondents' organizations survey results show that in a quantity range of 1–10 printers, 62% of respondents had local USB printers, 46% had network-based Multifunction Office Printers (MFPs), and all IT respondents had 3D printers. Furthermore, in a quantity range of 11–20 printers, 38% of respondents had network MFP printers, while 23% had local USB printers. We also have 8% and 15% of respondents with 41–100 network printers (MFPs) and local USB printers, respectively. This means that there is a potential security threat attack or exploitation that can occur with this quantity deployed.

Survey respondents were asked as well about their usage of print management software or deployment of cloud print infrastructure. Results show that 75% of IT respondents did not use or rely on Print Management Software or Cloud Print Infrastructure. While 11% use the solution. However, 14% of respondents are unsure whether their organization uses a print management solution.

Fig. 2 Printer vulnerability

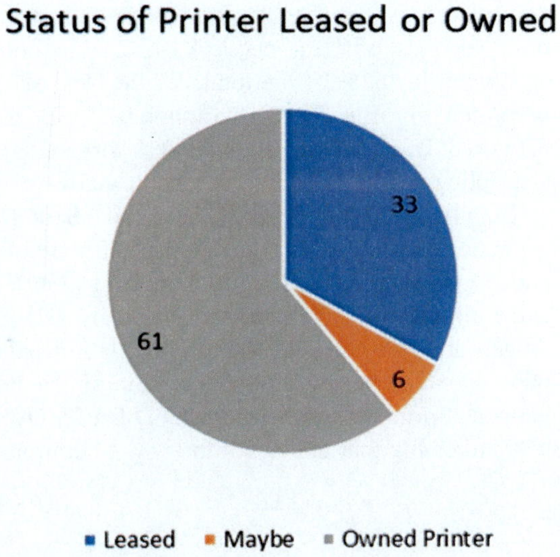

Fig. 3 Leased or owned printer

Figure 2 depicts the respondents' overall awareness of printer security threats and their general knowledge of it. According to the survey, 69% of respondents believe printers are vulnerable to security threats.

On the other hand, respondents' knowledge of their office printers, whether owned or leased, as shown in Fig. 3, shows that 61% of the devices were owned by the company, 33% were leased, and 6% were unaware of the device's status. To mitigate the risk of data loss or breach, having owned and leased devices during maintenance or repair necessitates the organization to have strong and clear data privacy policies for equipment. Moreover, as for the respondents' awareness of office printers, digital copiers, and 3D printers as endpoints, 92% of respondents agree, with the remaining 8% believing that these devices are not endpoints.

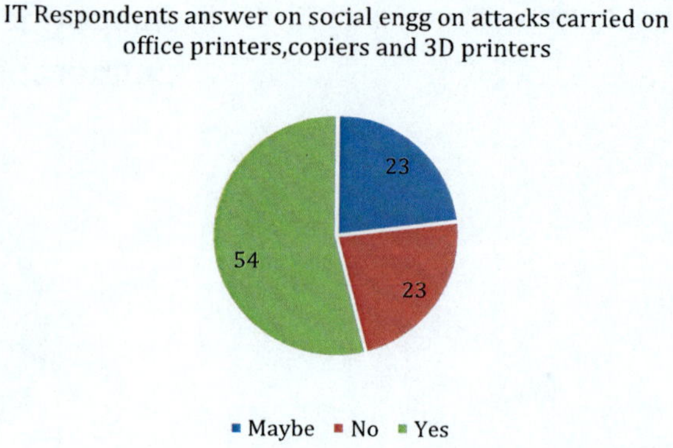

Fig. 4 Social engineering attack awareness

Figure 4 depicts the survey results regarding IT respondents' overall awareness and general knowledge of social engineering attacks on office printers. According to the results, 54% agree, 23% disagree, and 23% are unaware of social engineering attacks that can be used by cybercriminals to impose malicious actions on these devices. The survey results for IT respondents who allow different devices to use print services on their network show that only laptops and desktop computers accounted for 62% of the total. While 15% of IT responses indicate that users are permitted to print from smartphones, USB drives, and other devices on their network. This can put the network organization in jeopardy, especially if the smartphone and USB devices contain malware.

Finally, IT respondents were asked about their general knowledge or overall awareness of various attacks that can infiltrate office printers, MFPs, digital copiers, and 3D printers. According to the findings, 36% of IT respondents believe malware can infiltrate office printers, followed by DDoS at 18% and Man in the Middle Attack at 14%. Only 9% of the population believes that Cross-Site scripting and paper DoS can occur, while the remaining 9% believe that none of these things can happen to printers. Furthermore, LPD Buffer Overflow and "Printjack" attacks are at 0%, indicating that this is not the way to compromise their office printers.

3.3 Reconnaissance Simulation

An analysis of the potential consequences of raw 9100 port printing from "printjack" attacks was carried out further by the researcher using a free Shodan account, a search engine for the IoT, to determine how easy it would be to locate a printer's public IP address by searching for keywords like "port: 9100" or "printer." Passive reconnaissance is achieved by using Shodan as a technique to gather information about the targeted networks without actively engaging with them. This tool is also useful for finding specific devices running specific operating systems with specific ports open, in this case, port 9100. In addition, it can be used to find security vulnerabilities in printers. In addition, the researcher visits the MITRE Common

Vulnerabilities and Exposures (CVE) database and searches for exploits associated with printers. Besides the list, there are also the newest CVEs related to printers, which appear to be relevant to Shodan results. In the following simulations, the researchers exploited printers over the network. Thus, researchers were able to detect potential weaknesses in office printers for further analysis. For the passive reconnaissance and simulation of the penetration and security assessment test, the following tools were used: Shodan website, MITRE CVE Database Website, and ZenMap.

The simulation starts by creating an account to access the page of Shodan website. Passive reconnaissance is achieved by using Shodan as a technique to gather information about the targeted networks without actively engaging with them. This tool was also used to find specific devices running specific operating systems with specific ports open, in this case, port 9100. The result has shown the public IP address of the company and other information of the organization such as printer models, company location, and products. Security vulnerabilities in the printers were also mentioned in the findings. Then, the website of MITRE Common Vulnerabilities and Exposures (CVE) database was visited, and the results were displayed after searching for exploits related to printers. In addition to the list are the most recent CVEs related to printers that appear to be relevant to Shodan results. This study used printers over the network in the following simulations. As a result, researchers were able to identify potential flaws in office printers for further investigation.

4 Conclusions

An assessment was carried out in which IT professionals completed an online survey, and the results revealed that office printers are still vulnerable to social engineering attacks because they are viewed as nothing more than a tool for producing hard copies, and thus are the least prioritized when it comes to security implementation or protection of endpoints. The study confirmed and supported other related studies that office printers are frequently compromised due to open security ports on an office network, inside threats or lack of understanding from the IT personnel who happen to be the first layer of defense in a proactive approach to mitigate the probability of being a victim of cyberattacks. Data loss or data breach can happen either coming from an external or inside threat. With the result of survey response, it appears that IT practitioners are still not keen in considering or having a strong policy when it comes to data privacy from the said devices during maintenance and repair from service vendors. These open ports can be used to detect the device, operating system, and other information. Hackers can use services like Shodan to scan the Internet for network-connected devices and then target them. The reconnaissance phase is crucial to a successful attack, and Shodan only helps identify potential targets. It is easy for anyone with a computer to discover open ports on Internet-connected devices. Attackers exploit weak printer management to infiltrate organizations [13]. Few steps businesses may take to protect their printers are changing the default administrator username and password for printers and other endpoints or considering the device hardening checklist, maintaining a close eye on firmware updates of the office printers and updating them regularly, any printer settings that involve printing over the Internet should be disabled, disable all protocols except IP if they are not in use, limit scanning/copying/printing to the fewest number of subnets possible, and logging should be enabled, and logs should be reviewed periodically.

References

1. Giampaolo Bella, Pietro Biondi, and Stefano Bognanni. 2022. Multi-service threats: Attacking and protecting network printers and VoIP phones alike. Internet of Things, Volume 18. https://doi.org/10.1016/j.iot.2022.100507
2. Joy Reo. 2022. DDoS Hackers Using IoT Devices to Launch Attacks. https://www.corero.com/blog/ddos-hackers-using-iot-devices-to-launch-attacks/
3. Giampaolo Bella and Pietro Biondi. 2021. You Overtrust Your Printer. Springer International Publishing. https://doi.org/10.1007/978-3-030-26250-1_21
4. Bill Toulas. 2021. Researchers warn of severe risks from 'Printjack' printer attacks. Bleeping Computer. Retrieved from: https://www.bleepingcomputer.com/news/security/researchers-warn-of-severe-risks-from-printjack-printer-attacks/
5. Kevin Box. 2021. Printer Security Vulnerabilities and What You Can Do About It! Retrieved from: https://www.function-4.com/blog?p=printer-security-vulnerabilities-and-what-you-can-do-about-it-210810
6. Masha Kozinets. 2022. Printer Security and Vulnerabilities. Retrieved from: https://www.pharos.com/blog/printers-cybersecurity/
7. HP Development Company, L.P. 2021. https://h20195.www2.hp.com/v2/GetDocument.aspx?docname=4AA6-4525ENW
8. Emma Woollacott. 2021. HP printer vulnerabilities left enterprise networks open to abuse via 'cross-site printing' attack. Retrieved from: https://portswigger.net/daily-swig/hp-printer-vulnerabilities-left-enterprise-networks-open-to-abuse-via-cross-site-printing-nbsp-attack
9. Jason Firch. 2022. Common Types Of Network Security Vulnerabilities In 2022. Retrieved from: https://purplesec.us/common-network-vulnerabilities/#Vulnerability
10. Zoho Corporation. 2022. https://www.manageengine.com/network-configuration-manager/network-vulnerabilities.html
11. Brad Foster. 2022. Are Office Printers Vulnerable to Security Risks? Retrieved from: https://www.imageoneway.com/blog/printers-pose-security-risks
12. Jens Müller, Vladislav Mladenov, Juraj Somorovsky, and Jörg Schwenk. 2017. "SoK: Exploiting Network Printers," 2017 IEEE Symposium on Security and Privacy (SP), 2017, pp. 213–230, doi: https://doi.org/10.1109/SP.2017.47.
13. Grace Lorraine Diaz Intal, Delia Senoro, and Thelma Palaoag. 2021. User Experience Design for Disaster Management Mobile Application using Design Thinking Approach. In Proceedings of the 2020 4th International Conference on Software and e-Business (ICSEB '20). Association for Computing Machinery, New York, NY, USA, 7–13. https://doi.org/10.1145/3446569.3446587

Provisioning Deep Learning Inference on a Fog Computing Architecture

Patricia Simbaña, Alexis Soto, William Oñate, and Gustavo Caiza

1 Introduction

1.1 A Subsection Sample

Information digitization in IoT applications, such as services for smart cities, health, home, transportation, agriculture, etc., seems to be taking an important course when it refers to processing Cloud services in the Cloud itself. However, the flexibility required in digitized manufacturing for the Industrial Internet of Things (IIoT) depends on various aspects, such as architectural models employed at the edge of the plant, low latency, security, quality of service (QoS), communication links, energy storage and consumption [1], since the final devices should be read and controlled so that they operate in the times required at the process level, somewhat different than IoT [2]. This task may be solved with Cloud resources and the services offered by Internet providers. However, a scalable IIoT system would depend on an increase in its bandwidth and consequently on the costs that this involves; on the other hand, high latency is unavoidable in those countries that lack regional platforms of Cloud computing services, which is not appropriate for processes at the levels of control, planning and, even worse, for the execution of control and critical security [3, 4].

With this background, it is necessary that processes are performed at the edge of the plant, but without losing the potentiality provided by the Cloud, and one of the solutions under study is FC [5], since computing is performed close to the process, and its features for provisioning Cloud services are simple compared to virtual machines (VMs) of traditional hypervisors [6]. In a research taxonomy carried out

P. Simbaña · A. Soto · W. Oñate (✉) · G. Caiza
Universidad Politécnica Salesiana, Quito, Ecuador
e-mail: wonate@ups.edu.ec

L. Meng (ed.), *International Conference on Cloud Computing and Computer Network*, Signals and Communication Technology,
https://doi.org/10.1007/978-3-031-47100-1_6

in Civolani et al. [7], it is stated that a hybrid Fog/Cloud (F2C) collaboration might be the alternative to handle this situation, and the idea is focused on managing the services that will be used and executed in the Fog layer, through the control of such services, registering and storing them with the purpose of having a catalog of resources that will be optimally managed from a Platform as a Service (PaaS) platform. In the same course, Iram et al. [8] proposed the automation at the IoT level of the resources required by final devices, which would be possible through an environment as a service for private data centers or Kubernetes service, with the identification of the nodes performed by means of virtual network functions (VNF); despite this, the author specifies that the modules of the supervision engine and execution engine are not implemented, since this hybrid architecture is modeled from Cloudify. From a general point of view, the technology under consideration may fulfill various needs according to the requirements of the final client, such as edge computing, an effective platform for provisioning resources and meeting the low latency requirement; thus, Mouradian et al. [9] employed a large number of geographically distributed perimeter nodes and provisions a large number of pods; however, the authors state that its edge computing topology is simulated and emulated "from a real environment". Similarly, Toka [10] proposed an FC framework for orchestration known as FITOR, whose strategy is to optimize fog services provisioning (O-FSP) for IoT applications, improving aspects such as provisioning cost, use of resources, and acceptance rate; despite this description, the author states that there is a lack of practical implementations, but the experimental tests are simulated in an FIT/IoT-LAB and Grid5000 computing infrastructure.

The perspective of having various simulated works and, in some cases, emulating the real physical environment, is because it is required professionals proficient in IT and OT [11] for the technological development in the new industrial era; this situation still has a long way to go, typical of the paradigm currently going through.

This paper implements a Fog architecture constituted by a main node or Fog node (FN), for provisioning inferences from a DockerHub repository and orchestrating resources to balance the workload in the processing nodes known as mist nodes (MN), through Microk8s. In addition, Prometheus was used to perform a holistic tracking to collect CPU and RAM dynamic metrics during experimental tests, when there is a variation in the number of replicas to provision, orchestrate and execute DNN services for object recognition.

This document consists of Sects. 1 Introduction, 2 Methodology, 3 Analysis of Results and finally the conclusions in 4.

2 Methodology

Knowing the FC architectural framework and the works related to the provisioning and orchestration of services from the Cloud, a standard hierarchical architectural model consists of three layers [12], and thus an FC architecture was implemented as shown in Fig. 1.

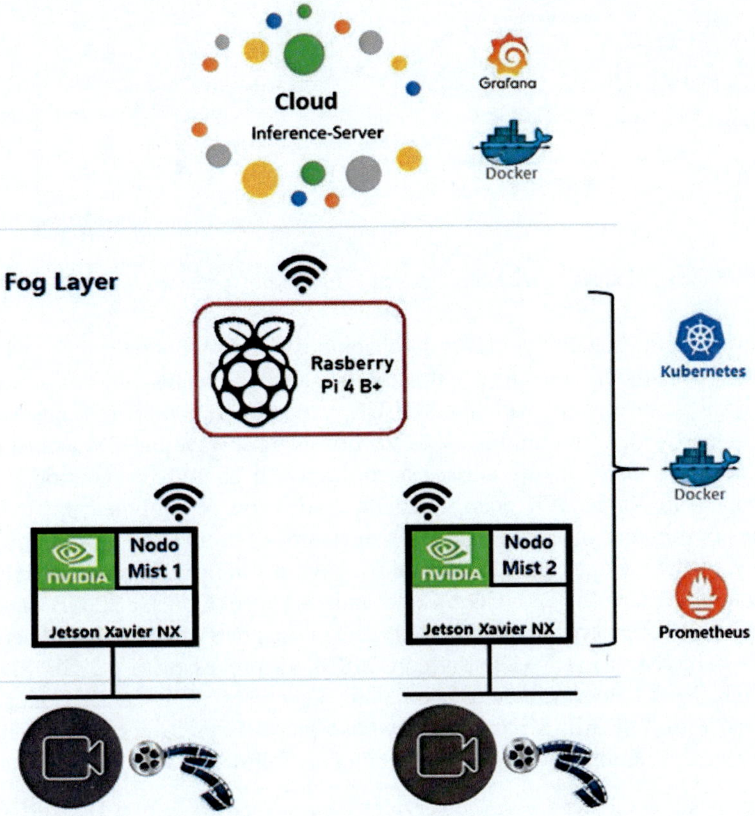

Fig. 1 FC architecture

A Raspberry Pi Camera Module 2 was placed in a job shop, to have available videos of a working environment in an I4.0 Laboratory; it is specified in Tang et al. [13] that the Cloud service provisioned in this document may be executed in file formats such as MPEG-4/AVC, MPEG-H, VP8, VP9, MPEG-2, MJPEG, and MPEG-4; the latter is the one employed for this case study.

The Fog layer is constituted by two edge devices on Nvidia boards and an FN on a Raspberry board. The AI models that are trained and present in the Nvidia inference server framework are imageNet for image recognition, segNet for semantic segmentation, poseNet for pose estimation, and detectNet for object detection. For this case study, a DNN model already trained with 91 classes of Jetson-Inference was used, its network architecture is SSD_MobileNet_V2 (see Fig. 2), which provides speed in object detection while maintaining the complexity of the network as decreases the vector spatial density.

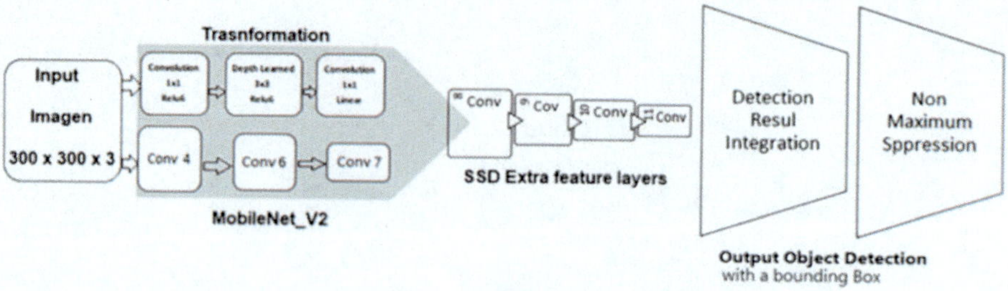

Fig. 2 FC SSD_MobileNet_V2 Object Detection framework. (Adapted from [14])

This general scheme presents two compositions: MobileNet V2, which is a network considered a feature extractor because it creates an image map with high-level features, and with the ReLU6 activation function, it limits scaling to high values with a maximum of six layers; in this way, the SSD object locator classifies the entries in the classes accordingly. It should be mentioned that the Nvidia Jetson Xavier NX boards include a software development kit (SDK) that enables optimizing inferences for AI when running from a GPU, obtaining a low latency in their execution [15, 16]; they have a 1900 MHz six-core ARM v8.2 processors, GPU with 384 CUDA Cores and 48 Tensor Cores, a 32 GB SD card and a 2.4 GHz Wi-Fi interface. On the other hand, a Raspberry Pi 4 B+ board performs as an FN, which has a 1.5 GHz six-core ARM v8 processor, a 32 GB SD card, a 2.4 GHz Wi-Fi interface, and IPv4 static addresses connected to an academic network with a 40 Mbps bandwidth, whose objective is to perform Gateway and Middleware functions and tasks, as specified as follows:

- It provisions Object Inference services from a repository in a DockerHub cloud platform, through containers that are light software packages [17].
- It orchestrates the service package or packages that were replicated, directing them to the MNs, balancing their load in CPU and RAM. It implements Kubernetes and deploys using the Microk8s software, compatible with Ubuntu 18.04 [18], i.e., through the creation of files with .yaml extension that contains declarative instructions [19].
- It is the Gateway between the Fog layer and the Cloud services through the IEEE 802.11/ac communication protocol, with a selected 2.4 GHz wireless channel.
- It enables monitoring the resources of the systems that constitute the Fog layer, collected through the Node Exporter and extracted by Prometheus for its visualization in Grafana; this was possible by enabling the 9100, 9090, and 3000 network ports, respectively [20].

Current manufacturing systems require the work of intelligent machines and operators, and this makes evident that these machines should carry out recognition of people and/or objects within the same environment, with response times appropriate for IIoT, thus avoiding regrettable events, for which various computing architecture technologies are being used to address this situation, as it was indicated in the introduction. Thus, through a study, dozens of replicas of deep learning inference services were executed by means of containers, which were provisioned from the cloud, for which two FC architectures were implemented; the first was developed for a vertical bidirectional communication, constituted by three levels (a mist-docker node, a fog-docker node and a cloud-docker service), and the second developed for a vertical and horizontal bidirectional communication, constituted by three levels (two mist-Kubernetes/docker nodes, one fog-Kubernetes/docker node and cloud-docker service), thus achieving a load balance in the edge nodes. The platform from which the DNN services were provisioned, and the software used in each node are described below:

- Docker Hub is the repository located in the cloud from which Docker provisions the image that encapsulates the necessary software [21], which is further orchestrated by Microk8s.
- Microk8s is an open-source system that enables the deployment of the main Kubernetes components, which makes possible orchestration between the processing MNs [22].
- Node Exporter is an exporter responsible for obtaining the dynamic metrics in the processing MNs and sending them to Prometheus in a compatible format.

3 Analysis of Results

The general resources of the system in the MNs were investigated through the implementation of the Prometheus exporter (Exporter Node) with the purpose of getting CPU and RAM metrics; this shows the percentage of CPU Load System, which depends on the number of processes waiting for execution. Thus, when it is about vertical bidirectional communication and with a single MN, Fig. 3 shows the corresponding behavior. Nevertheless, it should be considered the Compute Unified Device Architecture (CUDA) included in the Nvidia boards.

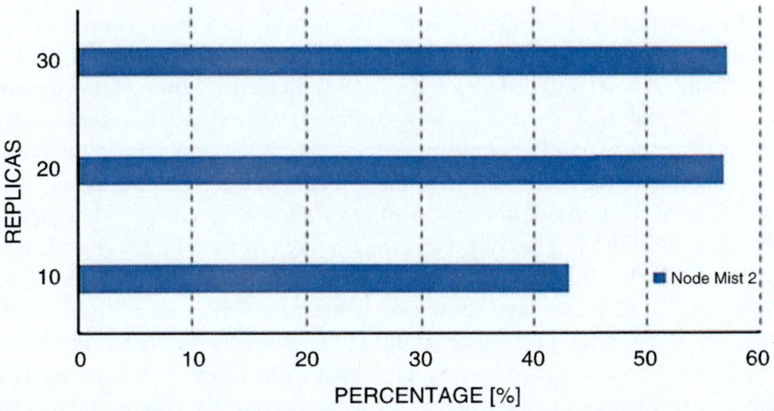

Fig. 3 CPU Load System consumption in an FC architecture with one MN

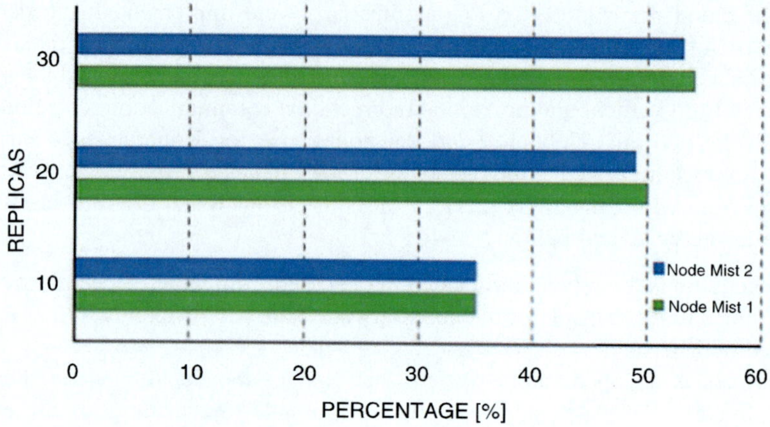

Fig. 4 CPU Load System consumption in an FC architecture with two MNs

It is evident that if the number of MNs increases, the computational effort of the nodes gets balanced, which is possible when the system is elastically scalable; for this purpose, Kubernetes is responsible for managing this scalability, as it is observed in Fig. 4. In this manner, the percentage value of the CPU Load System in each MN decreases, thus achieving a vertical and horizontal bidirectional architecture with a lower processing rate, and consequently, it provides a more efficient course of action to final devices. This situation is convenient if it is about a mobile robot that experiences an unforeseen event during its planned trajectory.

Figure 4 represents the metric of the mean load of the system during a period of 1 min, being able to observe that the architecture that orchestrates two MNs for balancing the workload yields a saving of 2.9 in processes to be executed, equivalent to 6.3% of the average load in each node; in other words, 60 deep learning inference services were provisioned, which were further orchestrated to the corresponding nodes, so that the DNN recognition programs of many objects per frame are executed and with a video resolution of 720 p for this case, as it is seen in Fig. 5.

During the execution of the AI algorithm for object detection, the CPU Busy metrics exhibited a second-order behavior, as indicated by the values presented in Table 1; this implies that, despite the long duration of the video, this does not have an impact on the amount of CPU consumption, which rather depends on the number of objects per frame regardless that the same object is detected n times.

Regardless of the number of replicas of DNNs to be executed, video time and/or number of NMs in the Fog architecture, the features of the Nvidia graphical boards housed in the NMs with respect to the dedicated VRAM memory show that, as expected, the RAM percentage remained constant at 40%, in contrast with what happens with the service provisioning and orchestration process carried out by the Raspberry board housed in the FN, which increases its RAM in 3% during 122 s for the case of a single node and during 57 s for the case of two nodes, results that are

Fig. 5 Execution of the AI-detectNet program

Table 1 CPU Busy metric in the NMs as a function of the number of objects detected

Video	Number of objects detected	CPU busy [%]
1	55	12.8
2	75	36.2
3	163	49.3

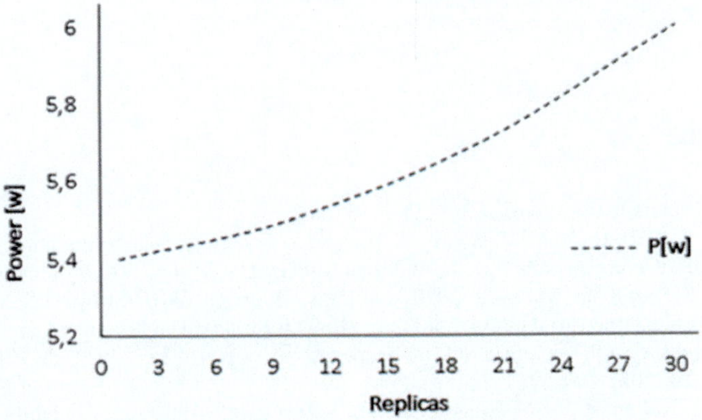

Fig. 6 Execution of the AI-detectNet program

independent of the execution processes of AI services in the MNs; on the other hand, this Fog device exhibits a similar behavior regarding power absorbed, with no difference regarding the number of operating MNs; however, this metric has a quadratic behavior as the number of provisioned and orchestrated dockers increases, with a determination coefficient R = 1, as it is observed in Fig. 6.

4 Conclusions

The FC architecture developed enables the communication between the FN and the MNs due to the use of Dockers that break the heterogeneity barriers between development boards, thus enabling that the containers are orchestrated toward the MNs, which guarantees management of the workload according to dynamic variables such as CPU and RAM, data that were extracted through Prometheus and visualized through Grafana for further analysis.

In the course of experimental assessments, we conducted a comparative analysis of the CPU Load System metric for Mist Nodes within a singular-node Fog Computing (FC) architecture, specifically designed for Flexible Cloud (FCFog) deployment. Since this is an elastically scalable architecture managed by Kubernetes, another MN was added, thus obtaining for the second architecture an improvement of 6.3% in each node with respect to the aforementioned metric, i.e., when a processing node is added, the number of processes to be executed per minute decreases and the performance level increases; in addition, taking into account the functions performed by the FN, it increases its RAM 3% in any of the architectures implemented, but the time of the provisioning and orchestration process improved 53.3%. Similarly, during this experimental test, the FN exhibited a quadratic behavior with respect to the power consumed, i.e., such power increased an average of 0.22 [w] for every 10 replicas.

When the DNN inference service is executed on videos with different duration times, the CPU Busy metric varies depending on the number of objects recognized per frame.

References

1. Author, F.: Article title. Journal 2(5), 99–110 (2016).
2. and D. V. I. A. D. D. D. A. Moskvin, "A Technique for Safely Transforming the Infrastructure of Industrial Control Systems to the Industrial Internet of Things," Autom. Control Comput. Sci., vol. 54, no. 8, pp. 841–849, 2020, doi: https://doi.org/10.3103/S0146411620080106.
3. V. R. Kebande, "Industrial internet of things (IIoT) forensics: The forgotten concept in the race towards industry 4.0," Forensic Sci. Int. Reports, vol. 5, p. 100257, 2022, doi: https://doi.org/10.1016/j.fsir.2022.100257.
4. J. Mustafa, K. Sandström, N. Ericsson, and L. Rizvanovic, "Analyzing availability and QoS of service-oriented cloud for industrial IoT applications," IEEE Int. Conf. Emerg. Technol. Fact. Autom. ETFA, vol. 2019-Septe, pp. 1403–1406, 2019, doi: https://doi.org/10.1109/ETFA.2019.8869274.
5. H. F. Atlam, R. J. Walters, y G. B. Wills, «Fog Computing and the Internet of Things: A Review», Big Data Cogn. Comput., vol. 2, n.o 2, Art. n.o 2, jun. 2018, doi: https://doi.org/10.3390/bdcc2020010.
6. R. Mahmud, K. Ramamohanarao, and R. Buyya, "Edge affinity-based management of applications in fog computing environments," UCC 2019 – Proc. 12th IEEE/ACM Int. Conf. Util. Cloud Comput., pp. 61–70, 2019, doi: https://doi.org/10.1145/3344341.3368795.

7. L. Civolani, G. Pierre, and P. Bellavista, "FogDocker: Start container now, fetch image later," UCC 2019 - Proc. 12th IEEE/ACM Int. Conf. Util. Cloud Comput., pp. 51–59, 2019, doi: https://doi.org/10.1145/3344341.3368811.
8. S. Iram, T. Fernando, and R. Hill, Connecting to smart cities : analyzing energy times series to visualize monthly electricity peak load in residential buildings, vol. 1, no. 880. Springer International Publishing, 2018.
9. C. Mouradian et al., "An IoT Platform-as-a-Service for NFV-Based Hybrid Cloud/Fog Systems," IEEE Internet Things J., vol. 7, no. 7, pp. 6102–6115, 2020, doi: https://doi.org/10.1109/JIOT.2020.2968235.
10. L. Toka, "Ultra-Reliable and Low-Latency Computing in the Edge with Kubernetes," J. Grid Comput., vol. 19, no. 3, 2021, doi: https://doi.org/10.1007/s10723-021-09573-z.
11. B. Donassolo, I. Fajjari, A. Legrand, and P. Mertikopoulos, "Demo: Fog Based Framework for IoT Service Orchestration," 2019 16th IEEE Annu. Consum. Commun. Netw. Conf. CCNC 2019, 2019, doi: https://doi.org/10.1109/CCNC.2019.8651852.
12. M. Lorenz, D. Küpper, M. Rüßmann, A. Heidemann, and A. Bause, "Time to Accelerate in the Race," The Boston Consulting Group, Boston, 2016.
13. B. Tang et al., «Incorporating Intelligence in Fog Computing for Big Data Analysis in Smart Cities», IEEE Trans. Ind. Inform., vol. 13, n.o 5, pp. 2140–2150, oct. 2017, doi: https://doi.org/10.1109/TII.2017.2679740.
14. Franklin, Dustin y Linderoth, Magnus, Deploying Deep Learning. 2022. Accedido: 27 de enero de 2022. [En línea]. Disponible en: https://github.com/dusty-nv/jetson-inference/blob/6bf94f753c727ea50f256fdec5fbe74bee540773/docs/aux-streaming.md
15. Ramalingam, B.; Elara Mohan, R.; Balakrishnan, S.; Elangovan, K.; Félix Gómez, B.; Pathmakumar, T.; Devarassu, M.; Mohan Rayaguru, M.; Baskar, C. sTetro-Deep Learning Powered Staircase Cleaning and Maintenance Reconfigurable Robot. Sensors 2021, 21, 6279. https://doi.org/10.3390/s21186279
16. «NVIDIA Jetson Xavier NX for Embedded & Edge Systems», NVIDIA. https://www.nvidia.com/es-la/autonomous-machines/embedded-systems/jetson-xavier-nx/ (accedido 2 de febrero de 2022).
17. NVIDIA Developer, «NVIDIA TensorRT», NVIDIA Developer, 5 de abril de 2016. https://developer.nvidia.com/tensorrt (accedido 27 de enero de 2022).
18. «What is a Container? | App Containerization | Docker». https://www.docker.com/resources/what-container (accedido 12 de febrero de 2022).
19. «MicroK8s vs k3s vs Minikube | MicroK8s», microk8s.io. http://microk8s.io (accedido 12 de febrero de 2022).
20. D. Santa Rendón, «Aplicación de una arquitectura basada en Service Mesh para una plataforma cognitiva utilizando Kubernetes e Istio», 2021, Accedido: 27 de enero de 2022. [En línea]. Disponible en: http://bibliotecadigital.udea.edu.co/handle/10495/20037
21. «Grafana basics», Grafana Labs. https://grafana.com/docs/grafana/latest/basics/ (accedido 12 de febrero de 2022).
22. B. Donassolo, I. Fajjari, A. Legrand, y P. Mertikopoulos, «Demo: Fog Based Framework for IoT Service Orchestration», en 2019 16th IEEE Annual Consumer Communications & Networking Conference (CCNC), Las Vegas, NV, USA, ene. 2019, pp. 1–2. doi: https://doi.org/10.1109/CCNC.2019.8651852.

A Comparative Analysis of VPN Applications and Their Security Capabilities Towards Security Issues

Eric B. Blancaflor, Jeremi An Armado, Christian James R. Cabral, Ezekiel Nathan B. Laurenio, and Jaystin Michael Joseph M. Salanguste

1 Introduction

1.1 Background of the Study

A virtual private network permits users from various networks publicly to establish a network connection that is secured. VPNs encrypt the Internet traffic you are presently using while concealing your online identity. Due to the longer time, it takes for real-time encryption to take place, third parties will have a hard time tracking your behavior in the online platform and snatching data. A VPN connection hides an individual's Internet data transmission and protects it from external illegal access. Unencrypted data may be viewed by anybody with network access and the motivation to analyze it, including hackers and Internet criminals. However, by using a VPN, cybercriminals are unable to decipher this data. In addition, a VPN enables safe data transfer for any remotely accessed data, such as data obtained through a company's network, and it also makes it possible to view information that is often only accessible in certain parts of the world [1]. The study focuses on the various VPN applications and their capabilities towards various security issues in the VPN and how effective the VPN application is towards that security issue. The study would discuss what are the possible implications of VPN applications in the Philippines and what could be the state of VPN usage in the country as well as what is the best VPN application that could be able to mitigate such security issues.

E. B. Blancaflor (✉) · J. A. Armado · C. J. R. Cabral · E. N. B. Laurenio · J. M. J. M. Salanguste
Mapua University, Makati, Philippines
e-mail: ebblancaflor@mapua.edu.ph

L. Meng (ed.), *International Conference on Cloud Computing and Computer Network*, Signals and Communication Technology,
https://doi.org/10.1007/978-3-031-47100-1_7

1.2 Significance of the Study

A comparative analysis of VPN applications and their security capabilities is important because it can help users understand the security features and limitations of different VPNs. This knowledge can be used to make informed decisions about which VPN to use and how to utilize it best to protect their data and online privacy. The significance of such a study is further increased by the fact that VPNs are commonly used to secure sensitive data, such as financial information and personal information, and to protect users from various online threats, such as hacking, surveillance, and identity theft. Therefore, understanding the security capabilities of VPNs can be crucial for individuals and organizations looking to safeguard their data and online activities.

There were also security vulnerabilities and threats for applications, and it would be difficult for any individual to take care of the situations [2]. There are countries that have security issues in the applications such as Lebanon where a study by Fadllalh et al. where they have revealed that there were 1645 dangerous vulnerabilities in the infrastructures and the whole country of Lebanon which are affecting its various sectors such as the field of technology and medicine. The 1645 vulnerabilities were due to the failure of patch management, a fundamental security best practice. At the time they were discovered to be vulnerable, all 1645 systems had a fix available from the manufacturer. The service will be affected even though there were patches and security updates involved and lack of patch management is the cause of the roots of the vulnerabilities [3]. This study also provides a key significance that even though an application deems it to be secured, there is still a chance that it will be affected, and a number of vulnerabilities will appear.

1.3 Objectives of the Study

This case study focuses on the VPN application's security capabilities and the challenges that may be involved with it. The study also will compare the security capabilities of different VPN applications and evaluate the effectiveness of VPN applications in protecting against security threats such as hacking, phishing, and malware attacks. The study will also identify any weaknesses or vulnerabilities in VPN applications that cybercriminals could exploit. Investigating in this study that the Philippines may be using a different standard from other countries in terms of their technology and security when it comes to VPN implementation, which is why this may be a downside in the long-run implementation of VPN, and it may be a downside in the long run implementation of VPN. However, the study could provide researchers with a solution for VPN implementation in the country as well as provide an advantageous and proper implementation of the technology.

2 Literature Review

2.1 Virtual Private Network Defined

Virtual private network is used as a function to create a private network on a public network, and how it will communicate, it will be encrypted. It is to establish a network tunnel with the help of encryption technology to protect the directional data transmission and create sniffing protection measures [4]. There are many ways for VPN configuration to happen within different platforms and systems, and most notables are within the server and client. In order to share services of the Intranet in public networks, configured servers are needed. Businesses mainly looked for performance, user experience, scalability, and security in a VPN service. A VPN solution must be able to improve throughput and latency performance as well as protect end-user productivity. It should also be scalable such that several users can simultaneously run on a single hardware platform without performance degrading. Detailed packet inspection, software filtering, and encryption ought to be attainable without degrading the effectiveness of the overall VPN system [5].

2.2 VPN User Experience

VPN user experience is vital for user productivity and satisfaction. A good VPN user experience is seamless and easy to use, with minimal setup and configuration. The user experience also includes features like automatic connection and disconnection and user-friendly management tools, reducing the need for IT support. In contrast, a bad VPN user experience may have connection issues and complex configurations that are difficult for those users who are illiterate in the technology or the technology in general. The experience users have with VPN is examined in the study by Nielsen and Gerdtsson. The study has looked into the experience of subjects with knowledge of VPN through focused interviews. There were seven interviewees, and they came from different backgrounds but mostly came from the IT industry. The interview results were subjected to different themes such as data collection, technicality, cyberattacks, and usage of VPN [6].

2.3 Security

Security is the most important aspect of VPN because it provides safeguarding the transferring of data, securing of IP and encryption of data as well as snooping. It also allows the user to browse the Internet through being anonymous and the anonymity of using VPN is great since it will mask your IP address. There are various strategies to check out and secure ports, packets, and network addresses that are solicited from incoming connections that will determine which traffic will be permissible on the

Table 1 VPN protocols

	Express VPN [9]	Nord VPN [10]	Surf Shark VPN [11]
IKEv2	✔	✔	✔
OpenVPN	✔	✔	✔
ShadowSocks			✔
Wireguard		✔	
No-Log Policy	✔	✔	✔
Ram-Only Server	✔	✔	✔
P2P Servers	✔		✔

network. Packet filtration is the most popular feature of a firewall where it inhibits the crossing of the router gateway from specifying the IP addresses. Authentication is another way that a VPN can help in solving Internet security issues where it will guarantee that the exchange between the users will be secured where it uses a shared key [7]. Moreover, the keys are subjected to hashing algorithms which provide a hash value and one key will be used to encrypt and decrypt the data that is being sent. Encryption is another way where it uses a private and public key [8].

In this study, comparisons of different safety protocols used in VPN were conducted. As shown in Table 1, express VPN has a compact user interface that can be simple to use. Bug Bounty programs are often done for finding errors in their service which ensures the security to be reliable. NordVPN offers the best UI which allows features to be browsed quickly and understandable security modules as the service offers a minimal design. SurfShark's UI is excellent for users who are not knowledgeable about the services. Moreover, it is equipped with IKEv2 and OpenVPN protocols. These protocols offer flexibility that uses an OpenVPN protocol via UDP or TCP whether it is for file transfer or streaming.

2.4 *Impact of Virtual Private Network*

With the rise of the VPN technology as well as certain current events that impact the usage of the said technologies, VPNs are deemed useful in creating impact when needed. During the period of COVID-19 pandemic, virtual private networks were used in various ways such as large enterprises with daily users ranging from 8000 to 80,000 from notable VPN solutions. The reason for this is remote work. While the physical world is at a pause from work, people are using the online world to work. There also was a lot of experimentation and research towards VPN implementations in the enterprise space due to the goal of protecting data from remote working. It came to a point that VPN providers were shipping equipment to clients before getting their orders for quick setups of the VPN. Even notable VPN providers such as Cisco offer free licenses that have a span of 13 weeks so setting up to scale up with their functions will be good and provide the clientele the time to purchase their product [12].

The impact of virtual private networks to society has been great towards providing the best possible experience for the user as well as companies to utilize their capabilities of securing and encrypting data in the network. Virtual private networks have provided countless solutions to various problems in the market to companies and during the pandemic, VPN is still being used since it has the capabilities for remote access through VPN where they could access certain files on the company's network. Moreover, the restrictions for the country at the height of the pandemic are not lenient, which is why companies have opted to provide the best possible environment for the company in accessing certain data from the network and have utilized the IT technologies to make their business operate at its full capacity even in these circumstances.

A study by Abhijith and Senthilvadivu where they focused on how the impact of VPN technology for the IT industries in the market during the time of the pandemic, most businesses desire an active-passive VPN configuration. It facilitates load balancing between two headend devices. In addition, businesses should be able to review the service level agreement they negotiated with their VPN service providers. Asking for an additional device from the hardware provider is preferable. Backing up each of the SLAs includes hardware as standard. Due to businesses implementing policies permitting work-from-home setups, VPN usage is currently experiencing a massive global increase rapidly to maintain activity throughout the pandemic. The ability of modern VPN providers to scale up to meet the increase in demand can be tested for the companies that are in need. However, it is also challenging for companies using such methods for first-time remote work setups since security requirements vary and are extensive. The best method for encrypting business data is with a VPN with the usage of home PCs to company networks, but prior to the pandemic, many companies had never really thought about implementing a VPN [12].

Big corporations were already utilizing proprietary and commercial VPN services. Such big corporations have seen a particularly strong demand for remote access VPNs. Although some of the remote access VPN expansions have resulted in greater business for VPN solution providers, others have been supported by the availability of a wide access to the customer's existing licenses. In other cases, businesses even used technology intended for those other projects that were designed for the VPN extension. Furthermore, Cisco provided customers with complimentary licenses during the pandemic for up to 13 weeks so they could grow exponentially and operate without difficulty while they had ample time to eventually continue with buying the product license. Pulse Secure provided customers with flexible and rapid licensing to support quick implementation. This made it possible for businesses to quickly adopt and benefit from Pulse Virtual Traffic Manager with Optimal Gateway Selection where the service will provide load balancing through the usage of virtual and cloud.

Virtual private networks are used in various ways. Cybersecurity personnel use the technology for sharing Intranet services with authorization and authentication within a public network. However, cybercriminals use the technology for identity spoofing and to be anonymous. There have been 479 vulnerabilities that were exposed and noticed in the public domain. However, only the top 28 vulnerabilities of the 479 have been exposed in 2020 alone, and it is because these vulnerabilities are in old VPN versions that are within the hardware. Another reason why there are constant vulnerabilities is because of the numerous hacking attempts towards VPNs, firmware has a hard time on consistent patches, therefore leading to many victims of VPN hacking. However, VPN networks are at high risk because hackers are noticing CVE and in order to exploit networks, they use auxiliaries and payloads [13].

2.5 Comparison of Virtual Private Networks for Business and User Usage

As shown in Table 2, SurfsharkVPN has the maximum devices compared to the other VPNs because surfsharkvpn has unlimited devices on a single account while expressvpn has five and nordvpn has six. Though Expressvpn leads on maximum devices, the system fell short of the speed of the other two VPNs by offering 75.12 Mbps speed on a 100 Mbps connection while Nordvpn offers 96.92 Mbps and Expressvpn 88 Mbps. On the other hand, ExpressVPN, NordVPN, and SurfsharkVPN have all shown friendly and fast responses to client questions and accommodations. The three VPNs offer three types of services for customers: email, 24/7 online chatting, and FAQ database. Based on client reviews, live chats are answered immediately with a maximum of 2 min response time, while email has a maximum of 42 h. The foremost function of VPNs is their layer of security for users' personal data, which expressVPN, NordVPN, and Surfshark VPN exhibit by applying the same standard encryption by the National Security Agency (NSA) for securing classified information, which is called AES or Advanced Encryption Standard with AES-256, also called 256-bit keys. All three VPNs make use of AES-256 to automatically protect clients' data. ExpressVPN and NordVPN use threat protection wherein the system defends data from trackers and malware while scanning and identifying threats before downloading files.

Table 2 Comparison of different VPNs premium

	ExpressVPN [9]	NordVPN [10]	SurfsharkVPN [11]
Convenience	88 Mbps on 100 Mbps connection	96.92 Mbps on 100 Mbps connection Meshnet feature	75.12 Mbps on 100 Mbps connection
Maximum device	Five devices	Six devices	Unlimited devices
Quality of service	FAQ database Email (ticketing system) 24/7 live chat	FAQ database Email (ticketing system) 24/7 live chat	FAQ database Email (ticketing system) 24/7 live chat
Security	AES-256 Threat protection	AES-256 Threat protection	AES-256

Table 3 Features comparison of the VPNs

	ExpressVPN [9]	NordVPN [10]	SurfsharkVPN [11]
Ad blocker	×	✓	✓
Bypasser	✓	✓	✓
Kill switch	✓	✓	✓
No activity logs	✓	✓	✓
Number of server countries	94+	50+	100+
Split tunneling	✓	✓	✓
24/7 customer support	✓	✓	✓

As presented in Table 3, ExpressVPN, NordVPN, and Surfshark all offer almost the same features, such as bypassing, kill switch, split tunneling, and blocking ads on websites during browsing. However, one of the VPNs differs from the other two VPNs, ExpressVPN, which does not provide any ad blocker features in their system. For a budget-friendly VPN, surf shark is the best option in terms of offering unlimited connection for devices at a cheaper price.

2.6 *Results and Discussion*

This section provides the survey results to 30 users of VPN applications. The survey was divided into three parts: Introduction to VPN, Familiarity with Virtual Private Networks, Experience, and feedback with VPNs where the results would provide an understanding of the VPN service and its applications.

In the results for questions regarding the Introduction to VPNs, the researchers wanted to showcase the respondents on whether they have used the service and its other market applications being used in the market. When answering the question, "What are some of the main reasons for using a VPN?"14 respondents (46.7%) are using VPNs to access blocked content in the country that they are in where in other parts of the world are accessible as compared to others. Seven respondents or 23.3% used it to be able to improve their online security while browsing the internet. Six respondents (20%) used it to protect their personal information. Finally, three respondents (10%) use VPNs to access restricted content while traveling – see Fig. 1.

In the results for questions about familiarity with virtual private networks, as shown in Fig. 2, it was to showcase the respondents' answers towards their familiarity with using the virtual private network application and its features. For the question "If you use a VPN, what VPN do you use from the list below and if others, please specify" 18 respondents (60%) are using NordVPN while eight respondents (26.7%) are using SurfShark as their VPN service of choice. Three of the respondents are using ExpressVPN for their VPN of choice. Lastly, one respondent said that they are using PureVPN as their VPN service of choice. For another question, "How comfortable are you with the level of security provided by your current VPN Provider", 22 respondents (73.3%) said that they were very comfortable with the

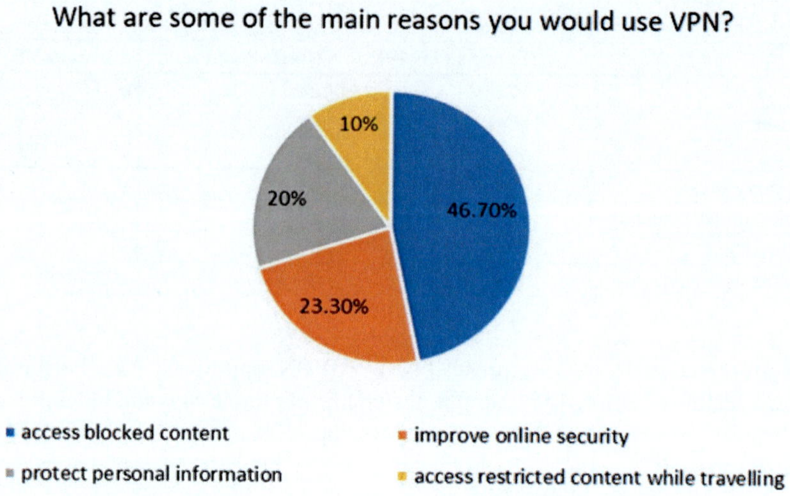

Fig. 1 A Pie Chart about the main reasons the respondents would use a VPN

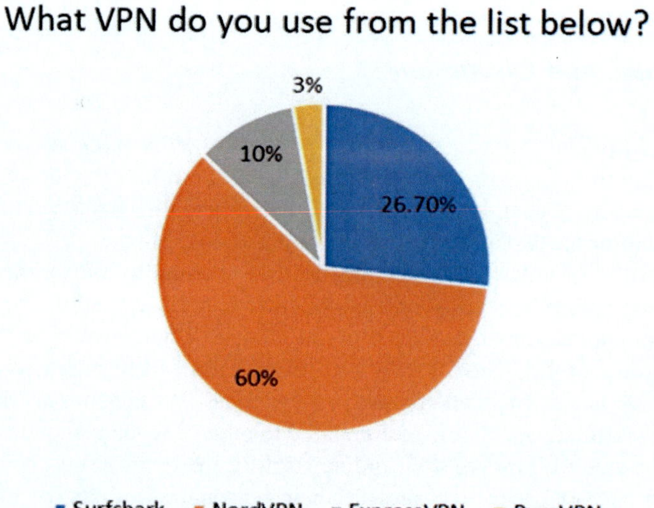

Fig. 2 A Pie Chart about what specified VPNs the respondents used

level of security that the VPN service provides. There are four respondents (13.3%) who are comfortable with the level of security that it entails. Lastly, there are also four respondents (13.3%) who are only neutral about the level of security that the VPN service provides.

For the final section of questions, the experiences and feedback with VPNs section showcases the respondents' answers about their experiences and feedback in using the VPN service and the applications that they have used. When answering

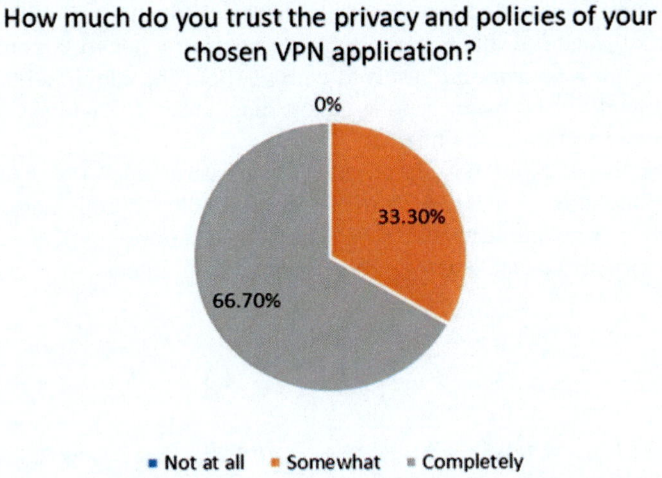

Fig. 3 A Pie chart about trust in the privacy and security policies of VPN

the question "Have you heard of VPN before", 30 respondents (100%) have heard the term VPN. When answering the second question "How familiar are you with the concept of using a VPN?", most of the respondents were very familiar with concept of VPN and 16 respondents were knowledgeable about the concept. Twenty-three percent of the respondents are familiar with the concept while 3% are neutral about the concepts where they might know some concepts but the entirety of it. Only one respondent answered as unfamiliar with the concept of VPN.

For the item, "How much do you trust the privacy and security policies of your top VPN application?", as shown in Fig. 3, 20 respondents (66.7%) said that they completely trust their VPN's privacy and security policies. However, 11 respondents (33.3%) somewhat trust their top VPN application's privacy and security policies. For the second question "Were there any specific features that you liked or disliked about different VPNs?", 60% of the respondents said there were no specific features they neither liked nor disliked. For the third and last question "Were there any issues or setbacks that you experienced while using different VPNs?", 90% of the respondents said they do not have any issues or setbacks while 10% said the issue was slow Internet.

3 Conclusion

In this study, VPN applications were studied for their effects on securing a network. Different VPN Providers that use different types of VPNs are considered in understanding how they work and how a network is affected by its absence. The capability of a VPN in securing a network and the vulnerability they make to the network is analyzed thoroughly to find out how they are made for the better. The advantages of using a VPN are commonly known for the significant effect they provide. This study proves that VPNs are for the better in terms of keeping a network secure. Security vulnerabilities of a network are resolved with a VPN. The additional layer secures data and network access by encrypting data, hiding a user's IP address, and providing a secure connection to the Internet.

The three VPNs, NordVPN, ExpressVPN, and Surfshark VPN, have the same Advanced Encryption Standard with AES-256, also called 256-bit keys for securing data. The chapter recommends that a VPN must have a no-log policy and operate a RAM-only server to create a network best secured from threats.

References

1. Kaspersky: What is VPN? How it Works, Types of VPN. https://www.kaspersky.com/resource-center/definitions/what-is-a-vpn, last accessed 2023/01/21.
2. Eric Blancaflor, Clarion Von Harvey Banzon, Craig James J Jackson, Jameela Nadine Jamena, Jeffrey Miraflores, and Leanne Kirsten Samala. 2021. Risk Assessments of Social Engineering Attacks and Set Controls in an Online Education Environment. In 2021 3rd International Conference on Modern Educational Technology (ICMET 2021). Association for Computing Machinery, New York, NY, USA, 69–74. https://doi.org/10.1145/3468978.3468990
3. Fadlallah, Y., Sbeiti, M., Hammoud, M., Nehme, M., Fadlallah, F.: On the Cyber Security of Lebanon: A Large-Scale Empirical Study of Critical Vulnerabilities. In: 2020 8th International Symposium on Digital Forensics and Security (ISDFS), pp. 1–6. IEEE, Beirut, Lebanon (2020).
4. Xu, Z., Ni, J.: Research on network security of VPN technology. In: 2020 International Conference on Information Science and Education (ICISE-IE), pp. 539–542. IEEE, Sanya China (2020).
5. S, A., Senthilvadivu, K.: Impact of VPN technology on IT industry during COVID-19 pandemic. In: International Journal of Engineering Applied Sciences and Technology, 2020, pp. 152–157. IJEAST, (2020).
6. Nielsen, E., Gerdtsson, M.: Qualitative analysis about the experience of VPN from people with software expertise in Sweden. In: Digitala Vetenskapliga Arkivet. Malmo University (2022).
7. Fortinet: VPN security: How secure is it & do you need one?. https://www.fortinet.com/resources/cyberglossary/are-vpns-safe, last accessed 2023/01/23.
8. Sharma, Y. K., Kaur, C.: The vital role of VPN in making secure connection over the internet world. In: IJRTE. Blue Eyes Intelligence Engineering & Sciences Publication, 2020.
9. ExpressVPN. ExpressVPN. Available: https://www.expressvpn.com/, last accessed 2023/01/23.
10. NordVPN. NordVPN. Available: https://nordvpn.com/, last accessed 2023/01/23.
11. Surfshark Surfshark. Available: https://surfshark.com/, last accessed 2023/01/23.
12. Abhijith M S and K. Senthilvadivu. 2020. Impact of VPN technology on IT industry during COVID-19 pandemic. International Journal of Engineering Applied Sciences and Technology 5, 5 (2020), 152–157. DOI:https://doi.org/10.33564/ijeast.2020.v05i05.027
13. Rama Bansode and Anup Girdhar. 2021. Common vulnerabilities exposed in VPN – a survey – iopscience. (2021). Retrieved January 23, 2023 from https://iopscience.iop.org/article/10.1088/1742-6596/1714/1/012045/meta

Improved Grey Wolf Optimization Algorithm Based on Logarithmic Inertia Weight

Xueying Luo and Lanyue Pi

1 Introduction

In recent years, with the continuous development of industry, agriculture, transportation, finance, national defense, and communication networks, the complexity of optimization in real life has risen by leaps and bounds. How to effectively solve these problems involving all aspects of life has become important for scholars to study. Particle Swarm Optimization (PSO) [1], Genetic Algorithm (GA) [2], Ant Colony Algorithm (ACO) [3], and other intelligent algorithms have been proposed to handle these difficulties.

Grey Wolf Optimization (GWO) was proposed by Australian scholar Seyedali Mirjalili et al. in 2014 [4, 5], which simulates the leadership and hunting mechanism of grey wolves in nature. It is easy to implement with a simple structure and few adjustment parameters [6]. Thanks to its good computational robustness and global search ability, the GWO algorithm has been widely used in image processing, image segmentation [7], flow shop scheduling [8], TSP [9], UAV three-dimensional route planning [10], and so on.

However, the GWO algorithm has some shortcomings, such as low solution accuracy and slow convergence. Especially when solving complex function optimization problems that are high-dimensional and multi-modal, it is easy to fall into local optimization. Based on the above defects, Mehak et al. [11] used chaos theory to improve the GWO algorithm and its global convergence speed. Using a dynamic evolutionary population algorithm, Saremi et al. [12] strengthened the local search ability of the GWO algorithm and accelerated its convergence speed to solve function optimization problems. Jayabarath et al. [13] improved the GWO algorithm

X. Luo (✉) · L. Pi
Zhengzhou Vocational College of Finance and Taxation, Zhengzhou, China

L. Meng (ed.), *International Conference on Cloud Computing and Computer Network*, Signals and Communication Technology,
https://doi.org/10.1007/978-3-031-47100-1_8

via the crossover algorithm and catastrophe algorithm, which tackled problems related to economic scheduling. Mittal et al. [14] promoted the performance of the GWO algorithm through a parameter adjustment strategy. Mohammad Sohrabi et al. [15] proposed an improved GWO to advance the quality of the final solution. Pin et al. [16] proposed an Alpha-guided GWO algorithm, which effectively guided the evolution of the population. A reference [17] proposed a random walk GWO algorithm to solve some defects of the wolf with leadership in GWO and improved leadership ability. Another reference [18] introduced the learning strategy based on elitist inversion and the simplex method into GWO, proposing a hybrid grey optimization algorithm based on elitist inversion. These improved strategies have promoted the performance of GWO to some extent from different aspects. However, some of them are more complex and their convergence speed fails to be increased significantly.

To simplify the improvement of GWO and promote its global search ability, convergence speed, and stability, an improved GWO algorithm based on logarithmic inertia weight is proposed in this paper.

2 Grey Wolf Optimization Algorithm

The steps of the GWO algorithm are as follows:

1. Initialize the population and set some parameters, including M as maximum iteration times, dim as dimension, N as population size, a, A, and C.
2. Calculate the fitness values of individuals in the population and record the three grey wolves α, β, and δ with the best fitness values.
3. Update the position of individuals in the population according to formulas (1, 2 and 3):

$$\begin{cases} D_\alpha = |C_1 \cdot X_\alpha - X| \\ D_\beta = |C_2 \cdot X_\beta - X| \\ D_\delta = |C_3 \cdot X_\delta - X| \end{cases} \tag{1}$$

$$\begin{cases} X_1 = X_\alpha - A_1 \cdot (D_\alpha) \\ X_2 = X_\beta - A_2 \cdot (D_\beta) \\ X_3 = X_\delta - A_3 \cdot (D_\delta) \end{cases} \tag{2}$$

$$X(t+1) = \frac{X_1 + X_2 + X_3}{3} \tag{3}$$

4. Update parameters a, A, and C according to formulas (4) and (5):

$$A = 2a \cdot r_1 - a \tag{4}$$

$$C = 2 \cdot r_2 \tag{5}$$

5. Calculate the fitness value of each grey wolf and update the positions of α, β, and δ.
6. Judge whether the end condition of the GWO algorithm is reached. If so, end the algorithm and output the optimal solution; otherwise, jump to step 3.

3 Improved Grey Wolf Optimization Algorithm Based on Logarithmic Inertia Weight

3.1 Logarithmic Inertia Weight Strategy

The updated formula of the logarithmic inertia weight strategy in this chapter is as follows:

$$y = -s \cdot \ln\left(\omega_{\text{start}} - (\omega_{\text{start}} - \omega_{\text{end}}) \cdot \frac{t}{t_{\max}}\right) + b \tag{6}$$

where $w_{\text{start}} = 0.9$ and $w_{\text{end}} = 0.2$, with the graph of logarithmic inertia weights shown in Fig. 1.

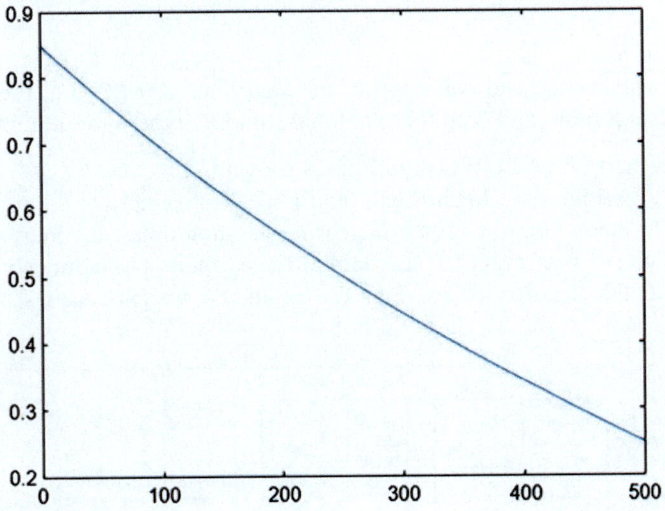

Fig. 1 Logarithmic inertia weight

As shown in Fig. 1, the logarithmic inertia weight is the largest at the initial stage of iteration, which is beneficial to expand the search range of the algorithm and avoid falling into the local optimization solution. Later, with the increasing iteration times of the algorithm, the curve begins to decrease downward, which is beneficial for the algorithm to obtain smaller values in a small range for a long time in the following iteration period, thus strengthening the local search ability of the algorithm and helping find the optimal solution in the local space.

3.2 *Improved Grey Wolf Optimization Algorithm Based on Logarithmic Inertia Weight*

The steps of the LGWO algorithm are as follows:

1. Initialize the population and set relevant parameters, including M as maximum iteration times, dim as dimension, N as population size, *a*, *A*, and *C*.
2. Calculate the fitness values of individuals in the population and record the three grey wolves α, β, and δ with the best fitness values.
3. Update the inertia weights according to the formula (6).
4. Update the position of individuals in the population according to the formulas (1), (7), and (3) in turn:

$$X_1 = \mathrm{s} \cdot X_\alpha - A \cdot D_\alpha, X_2 = \mathrm{s} \cdot X_\beta - A \cdot D_\beta, X_3 = \mathrm{s} \cdot X_\omega - A \cdot D_\omega \tag{7}$$

5. Update parameters *a*, *A*, and *C* according to formulas (4) and (5).
6. Calculate the fitness value of each grey wolf and update the positions of α, β, and δ;
7. Judge whether the end condition of the algorithm is reached at present. If so, end the algorithm and output the optimal solution; otherwise, jump to step 3.

The flowchart of the LGWO algorithm is shown in Fig. 2.

LGWO algorithm uses logarithmic inertia weight to improve that of the GWO algorithm. In other words, with the increasing iteration times, the inertia weight of the algorithm is maximum at the initial stage, then monotonously decreases according to the iteration times. Finally, the inertia weight reaches a minimum

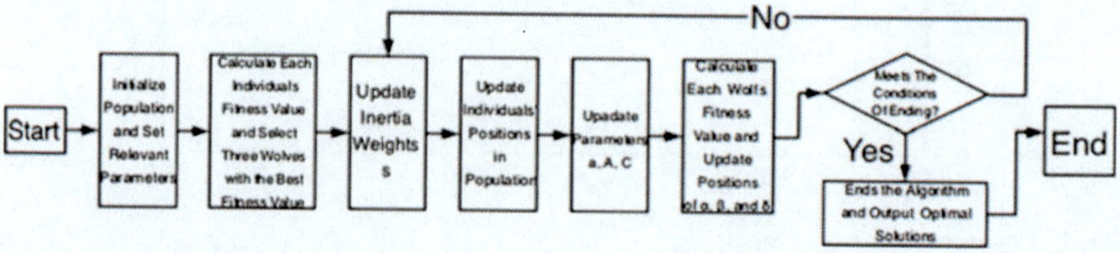

Fig. 2 Flow chart of LGWO algorithm

value. The larger inertia weight at the initial stage is beneficial to enlarge the search range of the algorithm and find the global optimal solution quickly. As the inertia weight decreases, the algorithm narrows the search range, and LGWO shifts from global optimum to a local optimum. Finally, this can help the algorithm find the global optimal solution more deeply in local space.

4 Simulation Experiment and Results

4.1 Experimental Design

The simulation environment is Windows 11, memory 8.00, MATLAB R2021a.

To verify the optimization ability of the LGWO algorithm accurately, six classical test functions are selected for simulation experiments, compared with five classical swarm intelligence optimization algorithms in low dimension (D = 30) and high dimension (D = 90, 300), respectively. The test function parameter settings used in this paper are shown in Table 1.where f_1–f_4 is a unimodal function and f_5 is a multimodal function. The dimensions of the five test functions are set to 30, 90, and 300, respectively. The parameters of the LGWO algorithm are set as follows. N = 50 as the population size and $t_{max} = 300$ as the maximum iteration times.

4.2 Analysis of Experimental Results

Comparison of Standard Deviation and Mean Value of Optimal Solutions of Six Algorithms

The comparison data of optimization results of five test functions are shown in Table 2.

Table 1 Test function parameter design table

Function number	Function expression	Dimension	Variable range value	Optimum value
F1	$f_1(x) = \sum_{n=1}^{\infty} (x_i^2)$	30,90,300	[−100,100]	0
F2	$f_2(x) = \sum_{i=1}^{n} \left(\sum_{j-1}^{i} x_j \right)^2$	30,90,300	[−100,100]	0
F3	$f_3(x) = \max_i\{\lvert x_i \rvert, 1 \le i \le n\}$	30,90,300	[−10,10]	0
F4	$f_6(x) = \sum_{i=1}^{n} \lvert x_i \rvert + \prod_{i=1}^{n} \lvert x_i \rvert$	30,90,300	[−10,10]	0
F5	$f_6(x) = \frac{1}{4000} \sum_{i=1}^{n} x_i^2 - \prod_{i=1}^{n} \cos\left(\frac{x_i}{\sqrt{i}}\right) + 1$	30,90,300	[−600,600]	0

Table 2 The mean and standard deviation of the optimal solutions of five algorithms in low dimension and high dimension

Test function	Algorithm	D = 30		D = 90		D = 300	
		Average	Standard deviation	Average	Standard deviation	Average	Standard deviation
F1	ALO	1.02E+04	4.01E+03	7.41E+04	1.18E+04	3.25E+05	5.08E+04
	FA	2.77E+03	1.92E+03	3.95E+04	1.42E+04	4.22E+05	4.57E+04
	POS	2.16E+03	374.9933	2.12E+04	2.71E+03	1.83E+05	1.78E+04
	FOA	2.91E+04	3.34E+03	1.36E+05	8.88E+03	6.03E+05	2.85E+04
	GWO	8.00E-04	6.00E-04	2.90	2.07	8.34E+02	2.72E+02
	LGWO	2.63E-124	3.97E-124	1.66E-119	1.25E-119	3.83E-116	3.69E-116
F2	ALO	4.02E+04	1.55E+04	3.50E+05	1.18E+05	3.09E+06	1.04E+06
	FA	3.06E+04	1.01E+04	3.15E+05	1.20E+05	3.48E+06	1.25E+06
	POS	1.20E+04	3.68E+03	1.37E+05	2.78E+04	1.42E+06	2.98E+05
	FOA	6.61E+04	1.67E+04	7.17E+05	1.99E+05	1.07E+07	4.01E+06
	GWO	2.21E+02	2.20E+02	2.23E+04	8.42E+03	4.69E+05	7.90E+04
	LGWO	1.50E-114	4.47E-114	1.61E-110	6.52E-110	7.48E-108	7.61E-108
F3	ALO	4.42	9.88E-01	5.80	4.86E-01	6.76	6.00E-01
	FA	7.37	1.03	9.51	2.05E-01	9.86	5.30E-02
	POS	9.15	3.66E-01	9.69	1.24E-01	9.91	2.60E-02
	FOA	6.28	2.47E-01	8.02	2.17E-01	8.90	1.17E-01
	GWO	8.93E-02	8.02E-02	3.10	7.44E-01	7.20	3.56E-01
	LGWO	9.69E-62	2.50E-61	6.09E-58	1.15E-57	8.95E-53	1.34E-52
F4	ALO	1.99E+02	4.03E+02	4.09E+27	1.27E+28	7.58E+143	2.35E+144
	FA	5.75E+01	1.62E+01	7.21E+20	1.78E+21	1.16E+123	5.18E+123
	POS	6.51E+12	8.69E+12	1.39E+48	5.88E+48	1.01E+171	Inf
	FOA	2.72E+06	6.72E+06	8.52E+31	3.77E+32	7.06E+122	3.16E+123
	GWO	3.20E-03	1.30E-03	5.88E-01	1.58E-01	2.34E+01	3.10
	LGWO	1.48E-63	1.45E-63	5.99E-61	5.10E-61	6.08E-59	2.76E-59

F5	ALO	8.58E+01	2.89E+01	6.27E+02	1.58E+02	3.06E+03	5.81E+02
	FA	1.19E+01	7.03	4.67E+02	1.28E+02	4.29E+03	5.12E+02
	POS	1.61	9.84E-02	1.01E+01	1.63	2.40E+02	2.11E+01
	FOA	2.81E+02	4.74E+01	1.29E+03	7.47E+01	5.37E+03	1.74E+02
	GWO	7.16E-02	5.62E-02	8.86E-01	1.25E-01	7.47	1.95
	LGWO	0	0	0	0	0	0
F1	ALO	1.02E+04	4.01E+03	7.41E+04	1.18E+04	3.25E+05	5.08E+04
	FA	2.77E+03	1.92E+03	3.95E+04	1.42E+04	4.22E+05	4.57E+04
	POS	2.16E+03	374.9933	2.12E+04	2.71E+03	1.83E+05	1.78E+04
	FOA	2.91E+04	3.34E+03	1.36E+05	8.88E+03	6.03E+05	2.85E+04
	GWO	8.00E-04	6.00E-04	2.90	2.07	8.34E+02	2.72E+02
	LGWO	2.63E-124	3.97E-124	1.66E-119	1.25E-119	3.83E-116	3.69E-116
F2	ALO	4.02E+04	1.55E+04	3.50E+05	1.18E+05	3.09E+06	1.04E+06
	FA	3.06E+04	1.01E+04	3.15E+05	1.20E+05	3.48E+06	1.25E+06
	POS	1.20E+04	3.68E+03	1.37E+05	2.78E+04	1.42E+06	2.98E+05
	FOA	6.61E+04	1.67E+04	7.17E+05	1.99E+05	1.07E+07	4.01E+06
	GWO	2.21E+02	2.20E+02	2.23E+04	8.42E+03	4.69E+05	7.90E+04
	LGWO	1.50E-114	4.47E-114	1.61E-110	6.52E-110	7.48E-108	7.61E-108
F3	ALO	4.42	9.88E-01	5.80	4.86E-01	6.76	6.00E-01
	FA	7.37	1.03	9.51	2.05E-01	9.86	5.30E-02
	POS	9.15	3.66E-01	9.69	1.24E-01	9.91	2.60E-02
	FOA	6.28	2.47E-01	8.02	2.17E-01	8.90	1.17E-01
	GWO	8.93E-02	8.02E-02	3.10	7.44E-01	7.20	3.56E-01
	LGWO	9.69E-62	2.50E-61	6.09E-58	1.15E-57	8.95E-53	1.34E-52

(continued)

Table 2 (continued)

Test function	Algorithm	D = 30		D = 90		D = 300	
		Average	Standard deviation	Average	Standard deviation	Average	Standard deviation
F4	ALO	1.99E+02	4.03E+02	4.09E+27	1.27E+28	7.58E+143	2.35E+144
	FA	5.75E+01	1.62E+01	7.21E+20	1.78E+21	1.16E+123	5.18E+123
	POS	6.51E+12	8.69E+12	1.39E+48	5.88E+48	1.01E+171	Inf
	FOA	2.72E+06	6.72E+06	8.52E+31	3.77E+32	7.06E+122	3.16E+123
	GWO	3.20E-03	1.30E-03	5.88E-01	1.58E-01	2.34E+01	3.10
	LGWO	1.48E-63	1.45E-63	5.99E-61	5.10E-61	6.08E-59	2.76E-59
F5	ALO	8.58E+01	2.89E+01	6.27E+02	1.58E+02	3.06E+03	5.81E+02
	FA	1.19E+01	7.03	4.67E+02	1.28E+02	4.29E+03	5.12E+02
	POS	1.61	9.84E-02	1.01E+01	1.63	2.40E+02	2.11E+01
	FOA	2.81E+02	4.74E+01	1.29E+03	7.47E+01	5.37E+03	1.74E+02
	GWO	7.16E-02	5.62E-02	8.86E-01	1.25E-01	7.47	1.95
	LGWO	0	0	0	0	0	0

Table 2 tests the mean and standard deviation of the optimal solutions of five algorithms in low dimension and high dimension. The average of the optimal solution represents their convergence accuracy, and the standard deviation represents their stability. It can be seen from Table 2 that the optimization effect of the LGWO algorithm is better than the other five algorithms, and its optimization result is also much higher than the other five. Under five test functions, the average of the optimal solution of LGWO is the closest to the global minimum value, which shows that LGWO has higher convergence accuracy and better optimization ability. Meanwhile, the standard deviation of the optimal solution of LGWO is smaller than that of the other five, indicating that the LGWO algorithm has higher stability.

Comparison of Convergence Curves of Six Algorithms

The convergence curves under 50 independent runs of LGWO, GWO, ALO, FA, POS, and FOA under 5 test functions are shown in Figs. 3, 4, 5, 6 and 7.

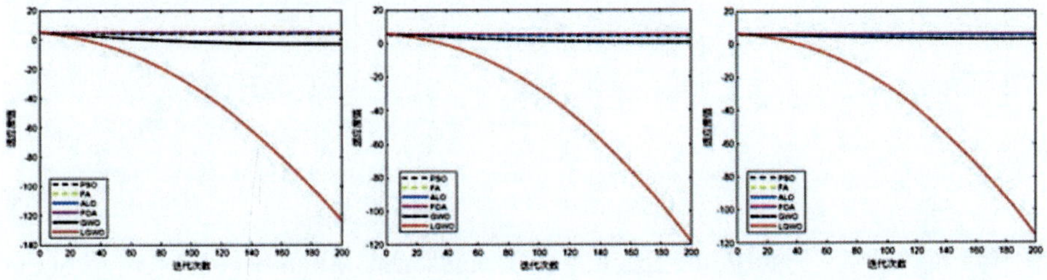

Fig. 3 Comparison of Average Convergence Curves of F1 (D = 30, D = 90, D = 300)

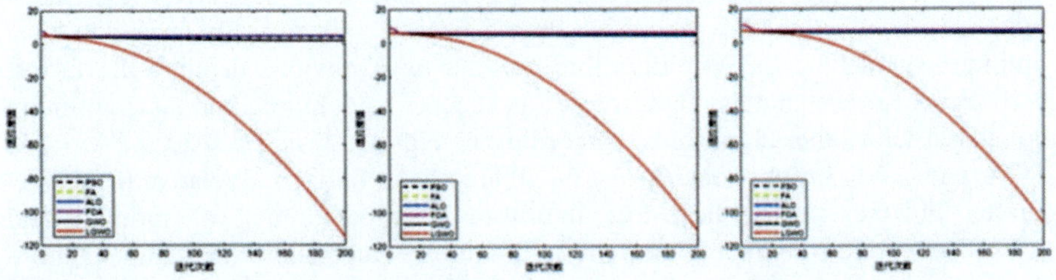

Fig. 4 Comparison of Average Convergence Curves of F2 (D = 30, D = 90, D = 300)

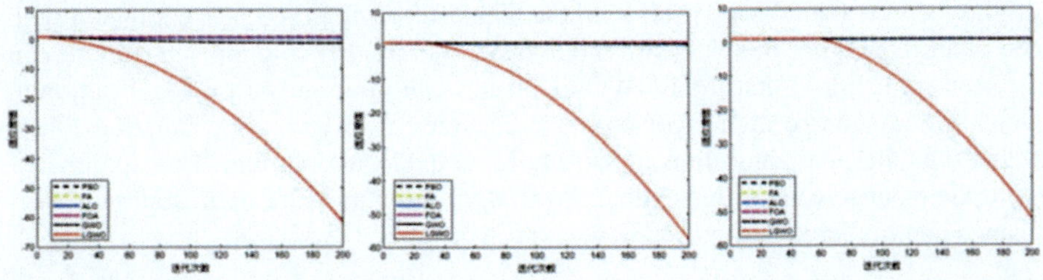

Fig. 5 Comparison of Average Convergence Curves of F3 (D = 30, D = 90, D = 300)

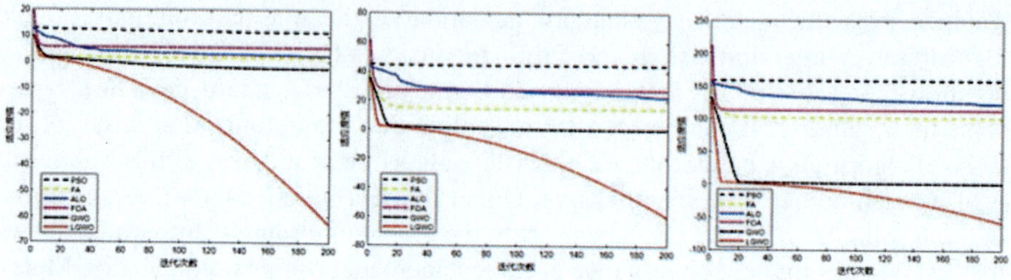

Fig. 6 Comparison of Average Convergence Curves of F4 (D = 30, D = 90, D = 300)

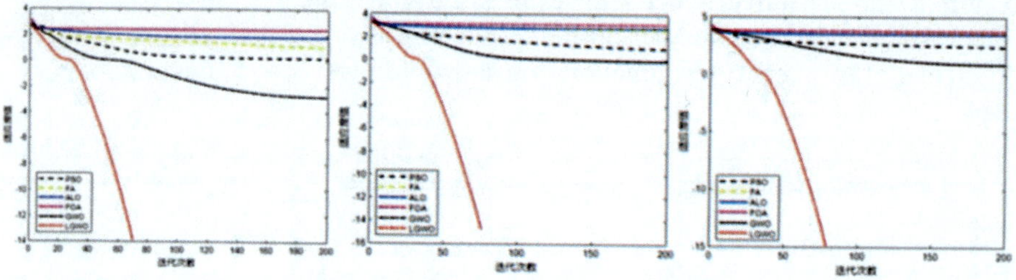

Fig. 7 Comparison of Average Convergence Curves of F5 (D = 30, D = 90, D = 300)

In Figs. 3, 4, 5, 6 and 7, the abscissa axis represents the number of iterations of the algorithm, and the ordinate axis represents the logarithmic average of the current generation population with optimal fitness value. According to Figs. 3, 4, 5, 6 and 7, with the increasing iteration times, ALO, FA, POS, and FOA have the weakest optimization effect, the lowest convergence accuracy, and are easily limited to local optimum, while the LGWO algorithm has the most obvious decline, the fastest convergence speed, and the highest solution accuracy, not falling into local optimum compared with other algorithms. According to Figs. 3, 4, and 5, ALO, FA, POS, FOA, and GWO in functions F_1, F_2, F_3 all tend to be horizontal relative to LGWO curves. In other words, these five algorithms soon fall into local optimum and LWGO continues to converge rapidly at this time, which shows that LGWO has a faster convergence speed and higher solution accuracy than the other five algorithms. As can be seen from Fig. 6, under the test of the function F_4, the convergence speed of LGWO in the early stage is the fastest. Although its convergence speed is slightly slower than that of itself with the increase of iteration times, compared with the other five algorithms, it is still much faster with higher solution accuracy. It can be seen from Fig. 7 that the LGWO algorithm can jump out of the local optimum quickly and achieve higher convergence accuracy with increasing iteration times, while the other five algorithms quickly fall into the local optimum. Thus, compared with the other five algorithms, the LGWO algorithm has better optimization performance and higher solution accuracy.

Time Complexity Analysis

The time complexity of the GWO algorithm is O(NMD), where N is the size of the grey wolf population, M is the number of iterations of the algorithm, and D is the dimension of the solution. The LGWO algorithm updates the inertia weights with formula 6 and updates the positions of three optimal grey wolves with formula 7. Compared with the position update formula of GWO in formula 2, LGWO essentially only adds a constant called inertia weight based on GWO. Therefore, LGWO does not increase the time complexity of the algorithm. To sum up, LGWO has faster convergence speed, higher solution accuracy, and more stability than GWO, with the same time complexity as GWO.

5 Conclusion and Prospect

In this chapter, LGWO is proposed to solve the problems of GWO, such as relatively slow convergence speed and the tendency to be limited to a local optimum. LGWO algorithm improves the inertia weight by using the characteristic of the logarithmic function, which not only expands the search range of the algorithm but also strengthens the local search ability of the algorithm. Simulation results show that LGWO has better optimization performance, faster convergence speed, higher convergence accuracy, and more stability than the other five algorithms.

Although the LGWO algorithm is improved in optimization performance, convergence speed, and convergence accuracy compared with the original GWO algorithm, it also has some limitations as follows: (1) Every calculation of the algorithm needs to preset the maximum iteration times in advance, which is difficult to predict the maximum iteration times of the algorithm in an industrial environment. Thus, adding an adaptive adjustment module to the GWO algorithm will be considered in the future, that is, giving inertia weight according to the fitness value of grey wolf or GWO with an adaptive inertia weight strategy. (2) This chapter studies the effectiveness of the logarithmic inertia weight strategy for improving the performance of the GWO but does not consider specific industrial applications. Future research will consider its industrial applications in reality, such as unmanned aerial vehicle (UAV) path planning, quantum information decision processing, and cancer diagnosis.

UAV path planning involves finding a safe, dynamically feasible, and optimal path for a drone to travel from its source point to a target point in a given environment through a series of algorithms, controls, and optimization methods. The path planning process involves four main components: (1) motion planning, which involves satisfying constraints such as flight paths and optimizing the path in terms of short path lengths and minimum turning angles; (2) trajectory optimization, which concerns optimizing the feasibility of the drone's motion during flight, including its speed, time, and path planning in terms of kinematics; (3) navigation, which combines motion planning, trajectory planning, collision avoidance, and positioning to provide overall control and monitoring of the drone's movement from one place to another; and (4) positioning, which involves understanding the drone's location so that human operators can make timely adjustments to its path in the event of unpredicted obstacles during its mission.

References

1. Kennedy, J. & Eberhart, R. C. (1995). Particle swarm optimization. Proceedings of IEEE International Conference on Neural Networks, 1942–1948.
2. Holland, J. H.(1975). Adaptation in natural and artificial systems. Ann Arbor: University of Michigan Press.
3. Colorni, A., Dorgo, M., Maniezzo, V, et al. (1991). Distributed optimization by ant colonies. Paris: Proceedings of European Conference on Artificial Life, 134–142.
4. Mirjalili, S., Mirjalili, S. M. & Lewis, A. (2014). Grey wolf optimizer. Advances in Engineering Software, 69:46–61.
5. Mirjalili, S., Mirjalili, S. M. et al. (2016). Multi-objective grey wolf optimizer: A novel algorithm for multi-criterion optimization. Expert System with Applications, 47: 106–119.
6. Holland, J. (1992). Genetic algorithms. Scientific American, 267(1): 66–72.
7. Wang, T. (2017). Research on intelligent image segmentation algorithm based on grey wolf optimization. Nanjing: Master Dissertation of Nanjing University of Posts and Telecommunications.
8. Lv, X. Q. & Liao, T. L. (2015). Permutation flow-shop scheduling based on the grey wolf optimizer. Journal of Wuhan University of Technology, 37(5), 111–116.
9. Xu, R. Q., Cao, M. & Huang, M. X. (2018). Research on the Quasi-TSP problem based on the improved grey wolf optimization algorithm: A case study of tourism. Geography and Geo-Information Science, 14–21.
10. Liu, C. A., Wang, X. P. , Liu, C. Y. & Wu, H. (2017). Three-dimensional route planning for unmanned aerial vehicle based on improved grey wolf optimizer algorithm. Journal of Huazhong University of Science and Technology (Natural Science Edition).
11. Mehak, K. & Sankalap, A. (2017). Chaotic grey wolf optimization algorithm for constrained optimization problems. Journal of Computation Design and Engineering. DOI:10.1010/j, jcde,2017.02.005.
12. Saremi, S., Mirjalili, S. Z. & Mirjalili, S. M. (2015). Evolutionary population dynamics and grey wolf optimizer. Neural Computing & Applications, 26(5), 1257–1263.
13. Jayabarathi, T., Raghunathan, T., Adarsh, B. R. et al. (2016). Economic dispatch using hybrid grey wolf optimizer. Energy, 111, 630–641.
14. Mittal, N., Singh, U., Sohi, B. S. (2016). Modified grey wolf optimizer for global engineering optimization. Applied Computational Intelligence and Soft Computing, 2016:8.
15. Nasrabadi, M. S., Sharafi, Y. & Tayari, M. (2016). A parallel grey wolf optimizer combined with opposition based learning. Bam, Iran: 2016 Conference on Swarm Intelligence and Evolutionary Computation (CSIEC), 18–23.
16. Pin, H., Chen, S. Y., Huang, H. X. et al. (2018). Alpha guided grey wolf optimizer and its application in two stage operational amplifier design. Changsha, China: 13th World Congress on Intelligent Control and Automation in 2018, 00.560–565.
17. Gupta, S. H. & Deep, K. S. (2018). A novel random walk grey wolf optimizer. Swarm and Evolutionary Computation. DOI:https://doi.org/10.1013/J.swevo.2018.01.001.
18. Emary, M., Zawbaa, H. M. & Grosan, C. (2018). Experienced grey wolf optimization through reinforcement learning and neural networks. IEEE Transaction on Neural Networks and Learning Systems, (29), 681–694.

Radio Frequency Identification Vulnerabilities: An Analysis on RFID-Related Physical Controls in an Infrastructure

Eric Blancaflor, Jed Ivan Fiedalan, Nicole Florence Magadan, Jhernika Mae Nuarin, and Ellize Angel Samson

1 Introduction

Radio frequency identification (RFID) technology is increasingly being used in many industries, demonstrating that it is a helpful tool for many companies, but there are challenges such as cost, lack of awareness, and security concerns. To protect RFID data and communications, strong security measures are necessary, including encryption and physical controls, as well as regular security audits and assessments [1]. RFID works by converting electrical impulses into radio waves, which readers and tags use to communicate with one another in an RFID system [2]. An RFID system's function is to transmit data from a portable device called a tag to an RFID reader to run a specific application based on the tag's identification or location data [3]. Compared to barcode technology, it offers far higher data integrity and accuracy, real-time response capabilities, and end-to-end visibility [4].

Identifying vulnerabilities in RFID-related physical controls is crucial for organizations to protect valuable assets, prevent system disruptions, comply with regulations and standards, and ensure the security, privacy, reliability, and integrity of RFID systems. Developing software-based security solutions for RFID-based security measures in IoT devices is challenging compared to other platforms due to the limitations imposed by the technology, such as speed, power consumption, cost, and high demand worldwide and creates a significant challenge for IT professionals in developing effective and efficient security solutions for IoT devices [5].

E. Blancaflor (✉) · J. I. Fiedalan · N. F. Magadan · J. M. Nuarin · E. A. Samson
School of Information Technology, Mapua University, Mapua, Philippines
e-mail: EBBlancaflor@mymail.mapua.edu.ph; jiffiedalan@mymail.mapua.edu.ph; nfjmagadan@mymail.mapua.edu.ph; jmrnuarin@mymail.mapua.edu.ph; eagsamson@mymail.mapua.edu.ph

L. Meng (ed.), *International Conference on Cloud Computing and Computer Network*, Signals and Communication Technology,
https://doi.org/10.1007/978-3-031-47100-1_9

This study provides an overview of RFID technology. This study aims to identify potential vulnerabilities in RFID-related physical controls in an infrastructure. This includes evaluating the effectiveness of encryption algorithms, frequency range and band, and RFID type. The study also identifies industry standards for employing RFID technology in an organization and to ensure interoperability of the devices. Furthermore, the study assessed the risks associated with identified vulnerabilities. The researchers would also like to raise awareness by providing a comprehensive comparative approach to different security mechanisms and protocols being employed in an RFID system. These objectives aim to protect the vital assets of an individual or an organization. Moreover, the scholars conducted a research survey identifying the knowledge of the public regarding RFID, common uses of RFID in the Philippines, and their confidence about the use of the technology.

2 Literature Review

2.1 RFID Tag, Reader, Antenna, Management Software, Action

RFID systems are composed of three basic components: RFID tags, RFID readers, and antennas. These components work together to transmit and receive data wirelessly.

Figure 1 shows the basic components of the RFID system. An RFID system is composed of three basic components: RFID tags, RFID readers, and antennas. RFID tags are small devices that contain a microchip and an antenna. The microchip stores

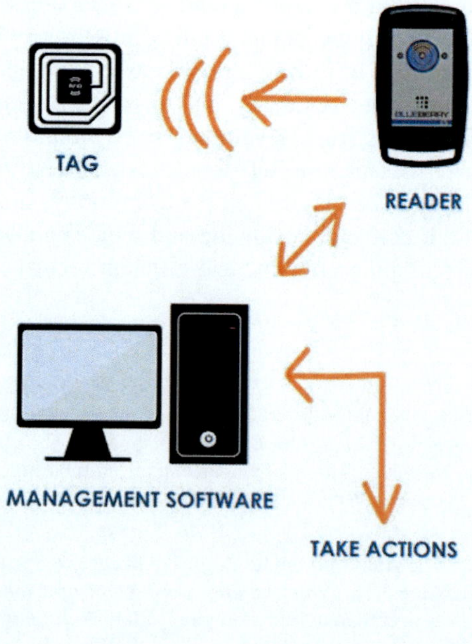

Fig. 1 Basic components of the RFID system [6]

Table 1 Contrast on passive, active, and semi-passive

RFID tag	Power source	Transmission	Vulnerability
Passive RFID	None	Dormant until initiated	Short reading range
Active RFID	Battery	Periodical	Limited battery life
Semi-passive RFID	Battery	Dormant until initiated	Limited battery life, but lasts longer than active tags

information, and the antenna allows the tag to communicate with the RFID reader. RFID tags can be passive, semi-passive, or active, depending on their power source. Passive tags do not have a power source, while semi-passive and active tags have their own power source. RFID reader is a device that sends out radio frequency (RF) signals to communicate with RFID tags. They receive data from the tags and send it to a host computer for processing. RFID readers can be handheld or fixed and can have different capabilities depending on the specific use case. The antenna allows the RFID tag to communicate with the RFID reader. The antenna receives the radio frequency (RF) signals from the reader and sends back the data stored in the microchip [6].

RFID systems utilize tags, readers, and antennas to wirelessly transmit and receive data [7]. Tags use radio waves to store and transmit data, and readers identify and track the object [8]. Passive, active, semi-active, and semi-passive tags exist [1, 4, 9]. Readers consist of antennas that emit and receive signals, a processing unit, a power source, a communication interface, memory, and a transmitter [2, 8, 10]. Antennas transmit and receive electromagnetic waves to communicate with tags, and the antenna's construction affects its performance [2, 4]. Management software manages and analyzes data collected from RFID readers and tags [1]. Actions can be triggered by the management system to control access or track attendance [11].

Table 1 shows the contrast between passive, active, and semi-passive RFIDs in terms of power source, method of transmission, and their respective vulnerabilities. The selection of an appropriate RFID tag category relies on the particular application scenario and criteria, such as the desired range of read, frequency of tag reads, and cost-effectiveness analysis.

2.2 RFID Layer Vulnerabilities, Attacks, and Security Measures

RFID (radio frequency identification) attacks are malicious attempts to gain unauthorized access to, or disrupt the normal operation of RFID systems [10]. These systems use radio waves to communicate between a reader and a tag, which can be attached to an object or person, in order to identify and track it. RFID attacks can target various aspects of these systems, including the physical components, communication protocols, software and applications, and business processes that rely on RFID data. RFID (radio frequency identification) attacks are classified into five layers: physical layer, network-transport layer, application layer, strategic layer, and multilayer attacks [12].

The physical layer includes all attacks that target the physical components of RFID systems, such as antennas, readers, and tags. Network-transport layer includes all attacks that target the communication protocols used by RFID systems. A permanent RFID tag disablement type of attack in the physical layer includes all potential hazards or threats that could result in the complete destruction or significantly reduced performance of an RFID tag. Physically removing or destroying an RFID tag could render it permanently unusable. Another attack in the physical layer is the relay attack. In this attack, an enemy serves as the man-in-the-middle in a relay attack. A reader and an antagonistic device are covertly inserted between two valid RFID tags. The radio transmission between a valid tag and reader can be intercepted and altered by this device. The hostile device then relays an ephemeral connection to the legitimate tag/reader from the legitimate tag/reader. The genuine tag and reader are persuaded to believe that they are speaking with one another directly. Separate devices, one for communication with the reader and one for communication with the RFID tag, might be utilized to make this kind of attack even more sophisticated [12].

Application layer includes all attacks that target the software and application that use RFID data. Examples include attacks employing unauthorized tag reading, modification of tag data, and attacks in the application middleware. Strategic layer includes all attacks that target the business processes and objectives of organizations that use RFID systems. Multilayer attacks include all attacks that combine elements from multiple layers, such as physical, network-transport layer, and strategic later attacks [12].

2.3 RFID Standards

There are several RFID standards and protocols that can be employed in an organization to ensure interoperability, compatibility, and security compliance between different RFID systems and devices – see Table 2. By using these RFID standards and protocols, organizations can ensure that their RFID systems and devices are secure, efficient, and productive. The International Organization for Standardization offers standards to provide interoperability and security between different manufacturers of RFID systems. Their purpose is to set standards in certain areas for radio-frequency-based identification systems. ISO 18000-x is one of their published works in this area [13]. These standards establish crucial information on specific existing layers and protocols to allow seamless transmission of data between two distinct manufactured devices. ISO 18046-x is employed in performance standards and ISO 18047-x is employed for compliance testing [13].

Table 2 RFID standards and security features [14]

Standards	Confidentiality	Integrity	Availability
EPC Class 0/0+	None	CRC error detection	High identification rate
EPC Class 1 Gen 1 and 2	None	CRC error detection	Lock/kill command
ISO/IEC 18000-2	No encryption	CRC error detection	None
ISO/IEC 18000-3	Dormant until the reader initializes transmission	CRC error detection	Different tag modes are available
ISO/IEC 11784-11785	Dormant until the reader initializes transmission	CRC error detection.	None
ISO/IEC 10536	Dormant until the reader initializes transmission	CRC error detection	Integrated with an anti-collision algorithm
ISO/IEC 15693	No encryption.	Error checking on the air interface	Password on lock command

EPC tags are a standard for RFID tagging in the supply chain. It is based on the EPCglobal Network, which is a global standard for identifying and tracking products using RFID technology. EPC class RFID tags are designed to work with the EPCglobal Network, which is a global standard for identifying and tracking products using RFID technology. EPC class RFID tags are small, wireless tags that can be attached to products or pallets. They contain a unique identification number (EPC) that can be read by an RFID reader. This number can be used to identify the product and track its movement through the supply chain. EPC class RFID tags can also store additional information such as product data, serial numbers, and expiration dates. ISO-Smart Cards have more security features including different encryption mechanisms such as 128-bit AES, triple-DES, and SHA-1 algorithms. These smart cards are often used and employed in the financial community as they offer more security and privacy features [14].

2.4 Research Survey

This study conducted a survey among students in the Philippines to test their awareness, knowledge, and confidence regarding RFID-enabled devices that are being employed in an organization, infrastructure, and institution. Schools and universities often use RFID systems on campuses for security, access control, and record monitoring. In the survey conducted by the researchers, thirty-five (35) respondents, all of whom are students at different schools in the Philippines. The survey questionnaire consists of close-ended questions with a combination of binary and 5-point scales.

Presented in formula 1 is the measurement of the subjective answers of the respondents with close-ended questions using a 5-point Likert scale.

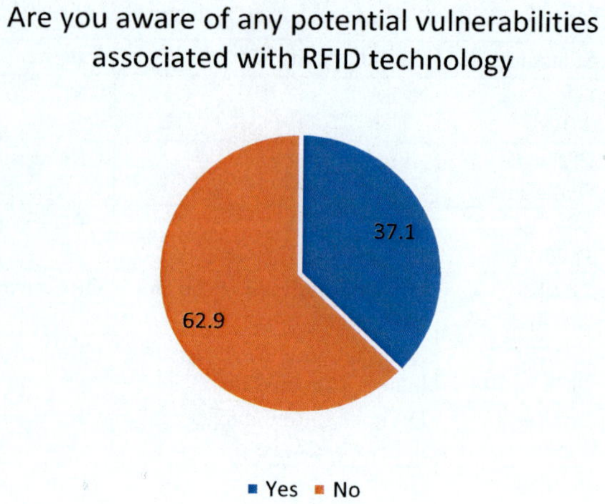

Fig. 2 RFID vulnerability awareness

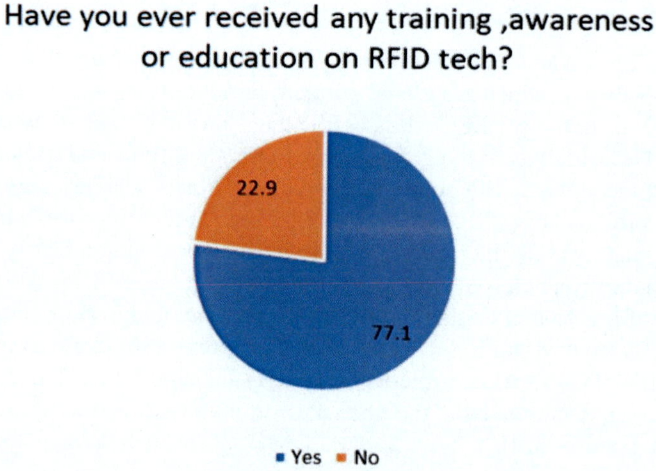

Fig. 3 RFID technology training and education awareness

$$N = (n1 \times 1) + (n2 \times 2) + (n3 \times 3) + (n4 \times 4) + (\times 5)/n \quad (1)$$

Figure 2 shows how many respondents are aware of the potential vulnerabilities using RFID-related physical control in a workplace or community based on respondents. Based on the results, 62.9% of the respondents are aware of the potential vulnerabilities with RFID systems as a physical security measure and 37.1% are not aware of potential vulnerabilities.

Figure 3 shows how many of the respondents were given training and education in RFID technology and its security implications. 71.1% of them answered "no", meaning 27 out of 35 of the respondents have not been given any training or

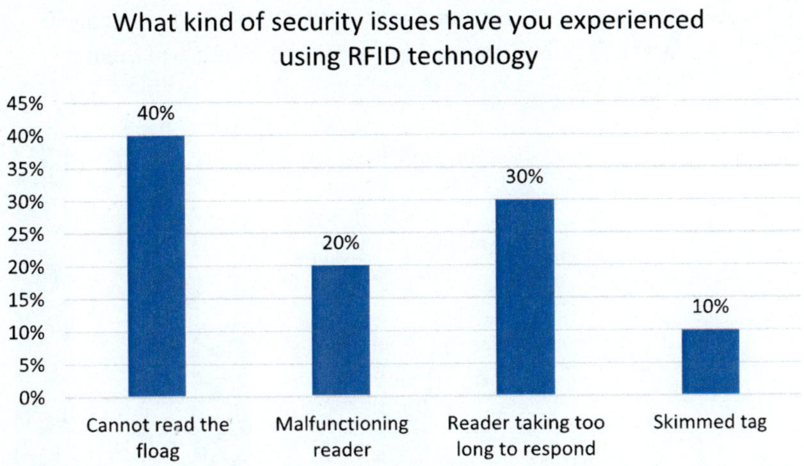

Fig. 4 RFID security issues

education regarding RFID. Moreover, 22.9% of the respondents answered "yes" or 8 out of 35 respondents were given at least a training or education regarding the technology.

As shown in Fig. 3, the confidence of respondents regarding Radio frequency identification (RFID) technology and its uses, results show that 40% of the respondents are aware and familiar with the technology and 11.4% do not have any confidence in Radio frequency identification (RFID) as a physical control of a workplace or community. On average, the respondents have neutral confidence in the use of RFID technology.

Figure 4 shows the issues experienced by respondents in using RFID technology as a physical security measure in an organization. The most common issue was the reader's inability to read the RFID tag, which was experienced by 4 out of 10 respondents. Three respondents reported that the RFID reader was taking too long to respond, while 2 others experienced a malfunctioning reader. One respondent reported an attack of a tag being skimmed.

Figure 5 shows the importance of RFID-related physical control in a workplace or community based on respondents. Based on the results, 57.1% of the respondents think that it is important to employ RFID systems as a physical security measure and 14.3% have neutral sentiments about it. The final score is 4.42, and, on average, the respondents think that it is important to use RFID technology in a workplace or community.

Presented in Fig. 6 is the familiarity of respondents regarding RFID technology and its uses. Based on the results, 37.1% of the respondents were very aware and familiar with the technology and 2.9% did not have an initial knowledge regarding it before taking the survey. On average, the respondents are familiar with the technology.

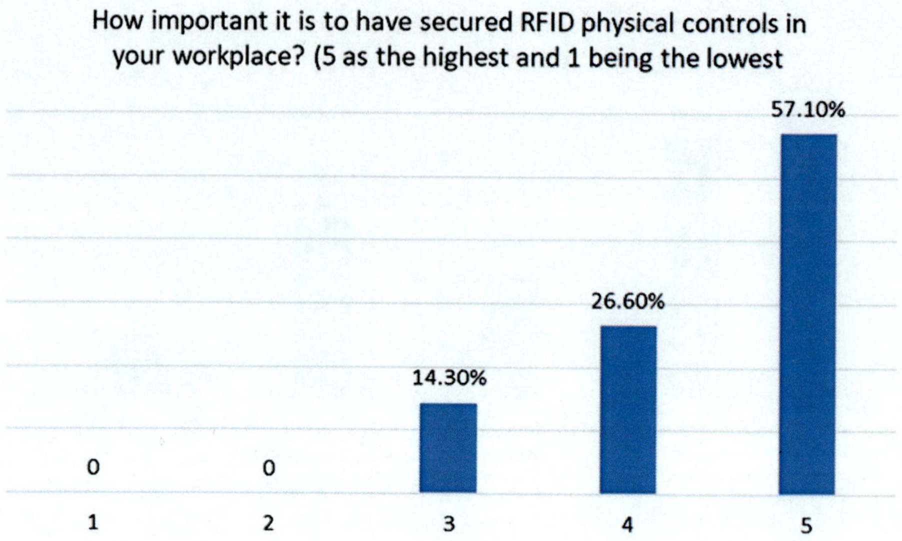

Fig. 5 Physical controls

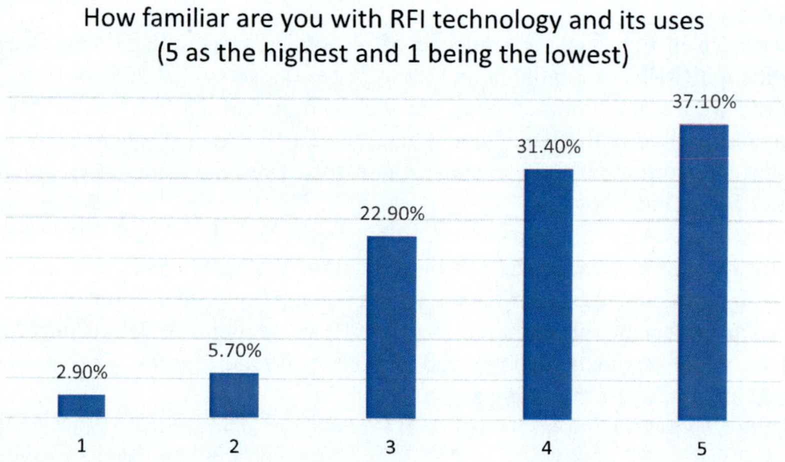

Fig. 6 RFID advantages

3 Conclusion

This study defined and identified the basic components of RFID systems and highlighted the several types and attributes of each device. The study also highlighted the vulnerabilities of using RFID technology as a physical control in an infrastructure. This research study has examined the vulnerabilities of RFID-related physical controls in an infrastructure, with the goal of identifying potential vulnerabilities, security and protocol, and standards and policies in employing RFID technology, and gathered insights by using a close-ended survey questionnaire to find out the knowledge, awareness, and confidence sentiments of the respondents. Based on the survey results conducted in this study, it shows the acceptable knowledge and awareness of the respondents with RFID-related security, still a need to continuously inform the users of Radio frequency identification (RFID) application of different threats [15] out there is a high recommendation for this study.

References

1. Ait, S., Lhadj Lamin, A. Raghib, and Abou El Majd, B.: Robust multi-objective optimization for solving the RFID Network Planning Problem, Mathematical Modeling and Computing, vol. 8, no. 4, pp. 616–626, https://doi.org/10.23939/mmc2021.04.616. (2021).
2. Ahsan, K., Hanifa, S., & Kingston, P.: RFID Applications: An Introductory and Exploratory Study. International Journal of Computer Science Issues, 7, 1–8 (2010).
3. Xiao, Q., Gibbons, T. and Lebru, H.: RFID Technology, Security Vulnerabilities, and Countermeasures. Supply Chain the Way to Flat Organisation, 4(4), Intech., doi:https://doi.org/10.5772/6668. (2009).
4. Williamson, A., Tsay, L.-S., Kateeb, I.A., and Burton, L.: Solutions for RFID Smart Tagged Card Security Vulnerabilities. AASRI Procedia, 4(4), 282–287. doi: https://doi.org/10.1016/j.aasri.2013.10.042. (2013).
5. Azuaje, R.: Securing IoT: Hardware Vs Software, Journal of Advances in Information Technology, vol. 9, no. 3, pp. 79–83. doi: https://doi.org/10.12720/jait.9.3.79-83. (2018).
6. Ait Lhadj Lamin, Samir & Abdelkader, Raghib & Abou El Majd, Badr. (2021). Robust multi-objective optimization for solving the RFID network planning problem. Mathematical Modeling and Computing, from https://doi.org/10.23939/mmc2021.04.616
7. Valero, E., Adán, A. and Cerrada, C.: Evolution of RFID Applications in Construction: A Literature Review. Sensors, 15(7), pp. 15988–16008. doi: https://doi.org/10.3390/s150715988. (2015).
8. Pateriya, R. K. and Sharma, S.: The evolution of RFID Security and Privacy: A Re-search Survey. In: International Conference on Communication Systems and Network Technologies, pp. 115–119. doi: https://doi.org/10.1109/CSNT.2011.31. (2011).
9. Guizani, S.: Implementation of an RFID relay attack countermeasure. In: 2015 International Wireless Communications and Mobile Computing Conference (IWCMC), pp. 518523. IEEE (2015).
10. Liu, B., Yang, B., and Su, X.: An improved two-way security authentication protocol for RFID system, Information, vol. 9, no. 4, p. 86, https://doi.org/10.3390/info9040086. (2018).
11. Morozova, T. V., Gurov, V. V.: Research in RFID vulnerability. In: Proceedings of the 2017 IEEE Conference of Russian Young Researchers in Electrical and Electronic Engineering (EIConRus), pp. 1–5. IEEE (2017).
12. Mitrokotsa, A., Rieback, M.R. & Tanenbaum, A.S. (2010). Classifying RFID attacks and defenses. Inf Syst Front 12, 491–505. https://doi.org/10.1007/s10796-009-9210-z
13. Guizani, S. (2015) "Implementation of an RFID relay attack countermeasure," International Wireless Communications and Mobile Computing Conference (IWCMC), Dubrovnik, Croatia, 2015, pp. 1318–1323, doi: https://doi.org/10.1109/IWCMC.2015.7289273.
14. Phillips, T., Karygiannis, T., & Kuhn, R. "Security standards for the RFID market," in IEEE Security & Privacy, vol. 3, no. 6, pp. 85–89, Nov.–Dec. 2005, doi: https://doi.org/10.1109/MSP.2005.157.
15. A. P. Abellon, C. J. Ariola, E. Blancaflor, A. K. Danao, D. Medel and M. Z. Santos, "Risk Assessments of Unattended Smart Contactless Cards," 2021 IEEE 8th International Conference on Industrial Engineering and Applications (ICIEA), Chengdu, China, 2021, pp. 338–341, doi: https://doi.org/10.1109/ICIEA52957.2021.9436788.

Part III
Computer Models and Artificial Intelligence Algorithms

Analysis of Bee Population and the Relationship with Time

Muyang Li, Xiaole Liu, Chen Qi, Lexuan Liu, and Kai Yang

1 Introduction

Honeybees are insects critical to every system on the planet [1–4]. Honeybees sustain the foundation of the food chain by pollinating and hence fertilizing numerous flowering plants. Bees are also essential to human survival and development, greatly contributing to agriculture, horticulture, food, and numerous other industries. According to the United Nations Ukraine [5], 35% of the world's agricultural acreage is dependent on these pollinators, which help produce 87 of the top 80 crops globally.

However, honeybee populations have been steadily declining since 2007 (United Nations Environmental Protection Agency), attributed to a variety of reasons, including both natural and human-caused hazards. The main threats to the long-term survival of bees, according to scientific studies, include climate change, habitat loss and fragmentation, invasive plants and bees, and limited genetic diversity. A more specific illustration is how the availability of floral resources decreases as more land is exploited. Therefore, bees must travel greater distances to meet their basic needs. The increased work will shorten their lifespan. As a result, the bee population has gradually decreased in many hives [6].

Considering the importance of bees and the severity of the current bee loss, many researchers have constructed models to investigate the correlation between the beehive population and a range of factors, aiming to restore bee colonies and increase bee populations. To study the impact of external factors (e.g., pesticide

M. Li · X. Liu · C. Qi · L. Liu
Amazingx Academy, Foshan, China

K. Yang (✉)
Sanya Science and Education Innovation Park, Wuhan University of Technology, Sanya, China

Xian Institute of Optics and Precision Mechanics of CAS, Xian, China

L. Meng (ed.), *International Conference on Cloud Computing and Computer Network*, Signals and Communication Technology,
https://doi.org/10.1007/978-3-031-47100-1_10

exposure, temperature, etc.) on bee populations, the literature [7] set up multiple experimental sites across the world and collected comprehensive field data, providing great insights for future researchers to refer to. The literature [8] discusses the impact of factors like nutrition, mortality, and health ratio on beehive populations and develops an integrated model to predict the growth trend of bee colony size with high accuracy. Other factors including but not limited to season, pollen type, diseases, and gene activities have been extensively studied [9–12].

This paper intends to provide a more comprehensive view of the topic by integrating and elaborating on the impacts of multiple factors influencing the population in bee colonies using appropriate mathematical models.

2 Models Overview

2.1 *Logic Flow of Model*

Over the past 23 years, the bee population has been on a steady decline. To tackle this problem, this chapter intends to investigate the change of bee population across five continents. Two models, a correlation model and an AHP-EWM model, are constructed to discuss the internal and external factors influencing the change of population in the bee colonies. With proper input, the correlation model outputs a percentage result to predict the population of bees. The AHP-EWM model is an integrated model to predict the bee population of a given beehive for a certain period of time. To verify the reliability of this integrated model, the result is compared with real experimental data.

2.2 *Data Approaching*

Applicable statistics are collected from secondary literature and reliable sites to populate the equations presented in this chapter. Some parameters change with conditions and therefore have different values in different countries and regions. Parameters with constant values were obtained from different sources, combined, and processed to better apply to our models.

A few variables used in the model were not readily accessible and were obtained by calculations and approximations. For the seasonal factor, the population of bees in each month is first calculated. Then, the average population in each season is obtained: 16006.58 bees in Spring, 44713.6 bees in Summer, 17479.56 bees in Autumn, and 5695.68 bees in Winter. The seasonal populations are then normalized with respect to the spring population to produce the coefficient of the seasonal indicators.

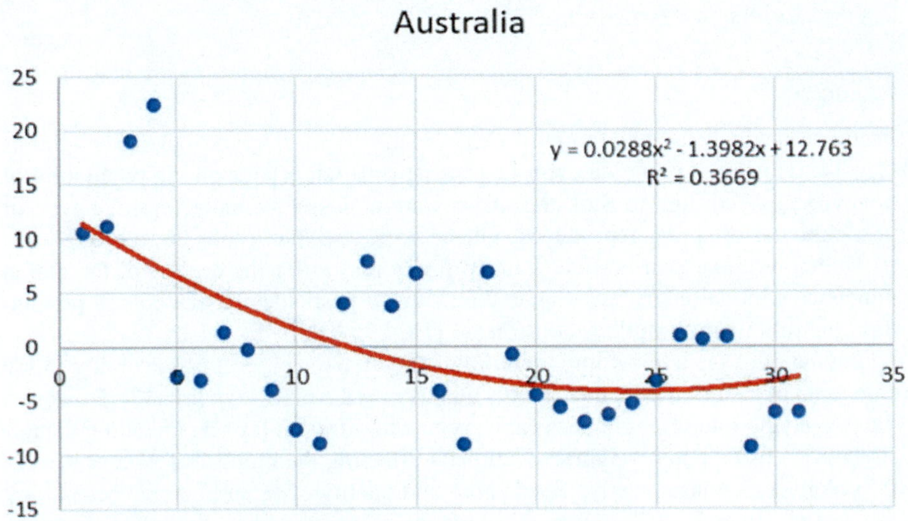

Fig. 1 Correlation of bees in Australia

2.3 A Primary Model for Problem 1

A model that considers population, time, and geographical locations, is constructed. The data on bee population over 30-year period from Europe, Australia, America, Asia, and Africa were collected. Figure 1 shows the regression model for each of these continents.

1. *Strengths and Weaknesses.*

The model discussed above has high accuracy and simplicity, and it is also capable of adapting to real scenarios. This is because the growth trends of the bee population in different continents have been modeled differently, which allows distinct population predictions to be made for different countries and regions and is, therefore, more applicable to and accurate in the real situations. Moreover, this model only considers the correlation between population and time, making the data processing and calculations relatively simple to make. Hence, this model is especially convenient when making a massive number of predictions of bee populations.

Despite its efficiency and applicability, this model lacks insights from different factors that may influence bee populations. Without these impacts, the reliability and flexibility of this model are doubtful. In view of this problem, some more comprehensive and in-depth approaches are explored in the following section.

2.4 Models for Problem 2

1. *Factors.*

Pesticide is widely used by farmers to kill, repel, or control harmful plants or animals [13, 14]. However, pesticide exposure has detrimental effects on the population of honeybees. According to Stokstad, if bees are exposed to imidacloprid, a type of pesticide, the offspring reproduced will be 75% less than those never exposed to it. Therefore, the use of pesticides is a primary reason for the decline of the global honeybee population. To investigate the factor of pesticides on bee colony population, the following formula is constructed [15, 16].

Laying eggs is the most important mission of the queen bee, as the ability to lay eggs is an indicator of both the queen's quality and the hive's quality [17]. However, the egg-laying rate of queen bees can vary due to different factors, including dietary structure, season, body weight, etc., directly affecting the number of worker bees in the colony and, consequently, the quantity and quality of an entire honeybee colony Hoan and Hoan [8]. Therefore, it is essential to find the correlation between egg-laying rate and brood population using appropriate mathematical functions.

$$P_{\text{egglaying}} = E \times H$$

The above formula is constructed to reflect the relationship between egg-laying rate (*E*), the possibility of successful hatching (*H*), and the percentage of bee population increased, predicting the future growth of a bee colony based on the egg-laying rate. The data is collected from Admin (2020).

2. *Basic Model 1 (Entropy Weight Model).*

EWM uses the characteristics of entropy to measure the degree of disorder of the system to determine the degree of dispersion of a factor. The greater the degree of dispersion of the factor, the greater the factor of the index on the comprehensive evaluation (that is, the weight). Therefore, the entropy weight method can be used as a tool to calculate the weights of each factor, which provides a basis for the comprehensive evaluation of multiple factors.

To build the model, since the data for the factors are not uniform, it is first necessary to take the absolute value of the data, that is $x_{ij} = |x_{ij}|$ and the processing method is as follows:

$$x_{ij} = \frac{x_{ij} - \min\{x_{1j}, \ldots x_{nj}\}}{\max\{x_{1j}, \ldots x_{nj}\} - \min\{x_{1j}, \ldots x_{nj}\}}$$

Then, use the equation below to individually calculate the weight, p_{ij}, of the ith scenario under the *j*th factor and the entropy index, e_j, of the jth factor

Table 1 Weight from EWM

Factors	Weight
Pesticide	0.0344
Egg laying rate	0.1363
Amino acid	0.1016
Infection	0.0808
Season	0.6469

Table 2 Uniformity check result

Maximal eigenvector	CI	RI	CR	Uniformity test result
5.075	0.019	1.120	0.017	Pass

$$P_{ij} = \frac{x_{ij}}{\sum_{i=1}^{n} x_{ij}} (j = 1, 2, 3, \ldots, m)$$

$$e_j = -k \sum_{i=1}^{n} p_{ij} \ln p_{ij}$$

with $k = \frac{1}{\ln(n)}$, satisfy $e_j \geq 0$.

Finally, using equation $g_j = 1 - e_j$, information entropy redundancy is calculated. Meanwhile, using formula $v_j = \frac{g_j}{\sum_{j=1}^{m} g_j}$ to get the final weight of each factor.

The result weights of factors by using EWM is shown in Table 1:

3. *Basic Model 2 (Analytic Hierarchy Process Model).*

AHP is one of the most used processes for analyzing models. During the process, the problem is split into three levels: the goal, the criteria, and the alternative.

After that, a comparison matrix has to be created, with the structure as follows:

$$C = \begin{bmatrix} c_{11} & c_{12} & c_{13} & \ldots & c_{1n} \\ c_{21} & c_{22} & c_{23} & \ldots & c_{2n} \\ c_{31} & c_{32} & c_{33} & \ldots & c_{3n} \\ \ldots & \ldots & \ldots & \ldots & \ldots \\ c_{n1} & c_{n2} & c_{n3} & \ldots & c_{nn} \end{bmatrix}$$

Among this matrix, c_{ij} represents the comparison of importance between c_i and c_j. In the article, the nine-scaling method is used to measure the importance of factors.

Then test the uniformity of the comparison matrix and calculate the index weight, using the incompatibility calculation formula *CI*:

$$CI = \frac{\lambda_{\max} - n}{n - 1}$$

The uniformity test result for AHP model is shown in Table 2.

Table 3 Weight from the integrated method

Factors	Weight
Pesticide	0.0535
Egg-laying rate	0.0316
Amino acid	0.0889
Infection	0.2084
Season	0.6177

4. *Extended Model (Optimal Weight Model).*

The weight factor V_K is a combined weight vector of V_0 and the kth evaluation method. Its deviation is $v_0 - v_k = (v_{10} - v_{1k}, v_{20} - v_{2k}, \ldots, v_{n0} - v_{nk}), k = 1, 2, 3, \cdots, s$. The target is to make the deviation of the combined weight vector and the s kind of weight as small as possible, and construct the optimization model in the sense of the least sum of squares of the deviations.

Finally, the composite score is calculated according to the following formula:

$$\text{score} = {V_0}^T x$$

Bring in the individual weight calculated in 4.1 and 4.2, the final weight of factors obtained are shown in Table 3.

To obtain a final percentage of the population against the initial population, W, for the evaluation of the population size of bees in relation to time. The following equations are performed: $W = -P_{\text{pesticide}} \times 0.05 + P_{\text{egglaying}} \times 0.03 + S \times 0.09 - P_{\text{infection}} \times 0.21 + \text{Season}\_c \times 0.62$ (Season_c for Spring is 1; Summer is 2.79; Autumn is 1.09; Winter is 0.35)

5. *Results Analysis.*

From our integrated model, it is concluded that the seasonal factor has the highest weight of 62%, followed by infection (21%), amino acid (9%), pesticide (5%), and lastly egg-laying rate (3.16%).

According to this modeling result, the change of seasons has a significant impact on the growth of the bee population. This result is consistent with findings by other studies that seasonal effects do have a strong correlation with bee population change. Infection, according to the literature [17–20], the change in the bee population is also of great significance, since if the disease is fatal, the impact will be devastating. This shows that this integrated model can accurately reflect the importance of different factors. Overall, most of the above results meet our expectations, proving the reliability of this model.

2.5 *Linear Programming Model for Problem 3*

To investigate how many beehives are required to pollinate a 20-acre cropland in 1 day, the following linear programming model is constructed. The logic flow of this model is shown in Fig. 2. This model explores the relationship between the bee

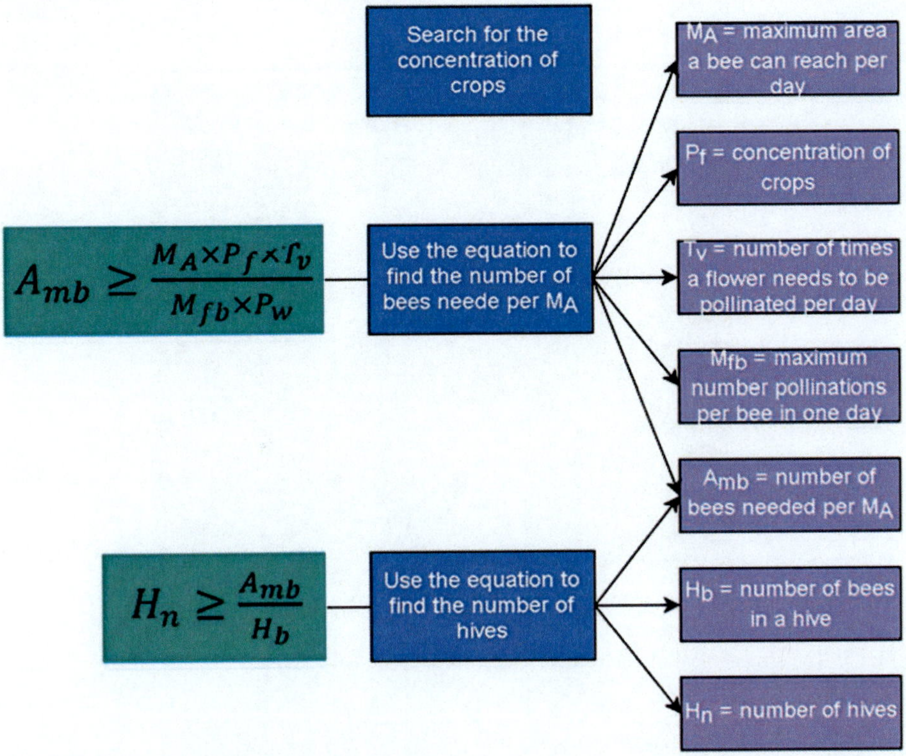

Fig. 2 Logic diagram of the solution to problem 3

population and the area of cropland by creating objective function and constraint conditions using data collected from credible websites and journals, which are presented as follows:

$$\begin{cases} A_{mb} \geq \dfrac{M_A \times P_f \times T_v}{M_{fb} \times P_w} \\ H_n \geq \dfrac{A_{mb}}{H_b} \end{cases}$$

In this part, our team predicted the number of beehives needed to pollinate a 20-acre land per day using the function above. First, we calculate the number of bees needed to pollinate a bee's reachable area (Amb) by this mathematical function.

$$A_{mb} \geq \frac{M_A \times P_f \times T_v}{M_{fb} \times P_w}$$

Here, the arrangement of cherry trees and fruit produced is shown in Fig. 3. Amb is measured by first multiplying the total number of pollinations a flower needs per day (Tv) by the number of flowers (Pf) and then times the mobile area of the bees

Fig. 3 The arrangement of cherry trees and fruit produced

(MA). This result is then divided by the number of pollinations a bee can do in 1 day (Mfb) times the proportion of foraging bees in a bee hive (Pw). A bee's moving space is within 6–20 km around its hive, as indicated by the MA. In this situation, the bees travel in a 20-acre (81,000 square meters) parcel of land, so their maximum mobile area (MA) is 20-acre.

Therefore, we multiply 1000 to change its unit from kilograms to grams since a cherry weighs approximately 10 g. Because one hectare is 2.47 acres, divide 2.47 to change its unit from hectares to acres. As 5 flowers can produce one cherry, the result is multiplied by 5. So, Pf equals $(6000 \times 1000 \div 10) \div 2.47 \times 5 = 1214574.9$ flowers per acre area. In that one flower will be pollinated at least 7 times for it to be successful.

1. *Strengths and Weaknesses.*

The above function is a general solution for multiple circumstances, as the land area, the concentration of crops and other parameters' values can be altered according to realistic actual situations. Moreover, the way we deal with this problem is to calculate a unit and multiply it by the corresponding coefficient. This problem-solving approach can be applied to complex problems that need to consider many factors. Besides, since the second function uses data from the previous model, as the value has already been evaluated through the process, the result is more scientific and reliable.

However, because the level of agricultural production in dissimilar countries is different, the quantity, quality, and intensity of crops also vary according to the production conditions. In the future, further investigation could involve presides classifications of agricultural production levels, and the concentration of crops should also have a complete set of measurement mechanisms as this is crucially related to the number of pollination bees.

Acknowledgments This work was in part supported by Project of Sanya Yazhou Bay Science and Technology City (Grant No: SCKJ-JYRC-2022-17) and Sanya Science and Education Innovation Park of Wuhan University of Technology (Grant No:2022KF0020).

References

1. Admin. (2020, May 9). How many eggs queen bee lays in a day? – Short-Fact. https://short-fact.com/how-many-eggs-queen-bee-lays-in-a-day/
2. Anita. (2022, May 29). How many cherries can a cherry tree produce? http://www.hnhuayukeji.com/15189.html
3. Antonio. (2019, November 5). How many cherries does a cherry tree produce? 0, 20, 50, 100, 400 kg? https://en.excelentesprecios.com/how-many-cherries-does-a-cherry-tree-produce
4. Betti, M. I., Wahl, L. M., & Zamir, M. (2014). Effects of Infection on Honey Bee Population Dynamics: A Model. PLOS ONE, 9(10), e110237. https://doi.org/10.1371/journal.pone.0110237
5. Botías, C., Hernández, R. M., Barrios, L., Meana, A., & Higes, M. (2013, April 10). Nosema spp. Infection and its negative effects on honey bees (Apis mellifera iberiensis) at the colony level | Veterinary Research | Full Text. https://veterinaryresearch.biomedcentral.com/articles/10.1186/1297-9716-44-25#:~:text=Under%20our%20experimental%20conditions%2C%20N.%20ceranae%20infection%20was,beekeeping%20profitability%20and%20have%20serious%20consequences%20on%20pollination.
6. Food and Agriculture rganization of United Nations. (2020, May 7). Land use in agriculture by the numbers | Sustainable Food and Agriculture | Food and Agriculture Organization of the United Nations. https://www.fao.org/sustainability/news/detail/en/c/1274219/
7. Greenwood, D. (2022). How Far Do Bees Travel? BeehiveHero [Internet]. https://beehivehero.com/how-far-do-bees-travel-from-their-hives.
8. Hoan, N. D., & Hoan, P. D. (2021). Factor effects on number of eggs laid of queen bee species Apis cerena at Northeastern region of Vietnam. https://lrrd.cipav.org.co/lrrd33/8/3399ndhoa.html
9. Jim, B. (2020, May 10). What is the size of a cherry? – KnowledgeBurrow.com. https://knowledgeburrow.com/what-is-the-size-of-a-cherry/#:~:text=One%20cherry%20weighs%205%20grams.%20How%20fast%20does,established%2C%20Cherry%20Blossom%20Trees%20require%20little%20care%20afterwards
10. Mull, A., Gunnell, J., Hansen, S., Ramirez, R., Walker, A., Lori Spears USU Department of Biology, & Zesiger, C. (2022, February). The conservation of bees: A global perspective | SpringerLink. https://link.springer.com/article/10.1051/apido/2009019
11. Pasquale, G. D., Salignon, M., Conte, Y. L., Belzunces, L. P., Decourtye, A., Kretzschmar, A., Suchail, S., Brunet, J.-L., & Alaux, C. (2013). Influence of Pollen Nutrition on Honey Bee Health: Do Pollen Quality and Diversity Matter? PLOS ONE, 8(8), e72016. https://doi.org/10.1371/journal.pone.0072016

12. Rittschof, C. C., & Robinson, G. E. (2013). Manipulation of colony environment modulates honey bee aggression and brain gene expression. Genes, Brain, and Behavior, 12(8), 10.1111/gbb.12087. https://doi.org/10.1111/gbb.12087
13. Stokstad, E. (2021, November 22). Pesticides can harm bees twice—As larvae and adults. https://www.science.org/content/article/pesticides-can-harm-bees-twice-larvae-and-adults
14. Switanek, M., Crailsheim, K., Truhetz, H., & Brodschneider, R. (2016, August 26). Modelling seasonal effects of temperature and precipitation on honey bee winter mortality in a temperate climate | Elsevier Enhanced Reader. https://doi.org/10.1016/j.scitotenv.2016.11.178
15. Torres, D. J., Ricoy, U. M., & Roybal, S. (2015). Modeling Honey Bee Populations. PLOS ONE, 10(7), e0130966. https://doi.org/10.1371/journal.pone.0130966
16. United Nations Environmental Protection Agency. (2020, October 26). Colony Collapse Disorder | US EPA. https://www.epa.gov/pollinator-protection/colony-collapse-disorder
17. United Nations Ukarine. (2020, May 20). We all depend on the survival of bees | United Nations in Ukraine. https://ukraine.un.org/en/112086-we-all-depend-survival-bees
18. Xiong. (2021, December 15). How many hours a day does a bee work? Qumifeng. https://www.qumifeng.com/mifengzhishi/5310.html
19. Yoko L, D., Nuno, C., Per, K., Joana, A., Jørgen A, A., Mette G, B., Marianne, B., Silvia, C., Julie, F., Geoff B, G., Annika S, J., Birgit, L.-K., Sara, L., Alice, P. M., da Silva Antonio, A., Beate, S., Peter Borgen, S., & José Paulo, S. (2021). Research project on field data collection for honey bee colony model evaluation. EFSA Supporting Publications, 18(7), 6695E. https://doi.org/10.2903/sp.efsa.2021.EN-6695
20. Yune, T. (2020, January 25). Why winter is a surprisingly crucial season for bees. Mic. https://www.mic.com/factor/how-honey-bee-populations-are-factored-in-winter-months-by-climate-change-19376661

Synthetic Speech Data Generation Using Generative Adversarial Networks

Michael Norval, Zenghui Wang, and Yanxia Sun

1 Introduction

Communication is imperative to the human species. Without communication, there would be a complete breakdown. One communication method is off-course speech. Speech is generated by the human vocal cords and other human anatomy systems working together. The research problem is to generate synthetic data and make the sound human-like and realistic. The model must be able to generate speech that addresses these voice synthesis limitations: naturalness, prosody, spontaneity, and ambiguities [1, 2, 16]. Training deep learning models for speech and emotion recognition data is needed. Qualitative data can be collected and extracted; however, if it could be generated artificially, it would be very beneficial. In order to generate synthetic audio clips, speech synthesis is used. Older legacy methods do not suffice because of the quality of the audio clips as well as the lack of emotion in the uttered speech. Deep learning speech models have huge commercial applications, especially for text-to-speech, speech recognition, and emotion recognition. More data allows for more training and subsequently better accuracy of deep learning models. The rest of this paper is organized as follows: Sect. 2 elaborates on text-to-speech techniques. This includes GAN and the use of the Tacotron 2 model to generate speech. Section 3 looks at the evaluation and overall settings of the architecture. Finally, Sect. 4 provides the conclusion of the paper.

M. Norval · Z. Wang (✉)
Department of Electrical Engineering, University of South Africa, Florida, South Africa

Y. Sun
Department of Electrical and Electronics Science Engineering, University of Johannesburg, Johannesburg, South Africa

L. Meng (ed.), *International Conference on Cloud Computing and Computer Network*, Signals and Communication Technology,
https://doi.org/10.1007/978-3-031-47100-1_11

2 Text-to-Speech Synthesis

Text-to-speech or speech synthesis is a known technology that was developed in the 1960s. Pre-recorder phrases were stored and recalled when required [3]. The latest methods involve deep learning and conditional generative adversarial network models [4]. Tacotron version 2 will be used in this paper [5, 6]. The speech corpus that will be used is the NCHLT Afrikaans Speech Corpus comprising 56 h of speech produced by eight authors. The corpus is licensed as creative commons and free to use [7, 8].

2.1 Legacy Synthesis

Computer-based speech-synthesis systems were invented in the late 1950s. In the beginning, linear predictive coding was used. Later this evolved into line spectral pairs. In 1980, an line spectrum pair (LSP)-based speech synthesizer chip was developed [3]. Most synthesized voices were male with AT&T creating the first female voice. Various techniques and technologies are used for speech synthesis. These include Concatenation synthesis, Unit selection synthesis, Diphone synthesis, Domain-specific synthesis, Formant synthesis, Articulatory synthesis, Statistical Parametric synthesis, and Sinewave synthesis [9, 12].

2.2 Deep Learning Speech Synthesis

Nowadays deep learning-based synthesis is used to produce a spectrum vocoder that translates text to speech. The most popular type of deep learning network used is called the generative adversarial network.

2.3 GAN

The basic architecture can be seen in Fig. 1. A GAN model consists of two sub-networks: A Discriminator network (D) and a Generator network (G). The generator network attempts to map a simple distribution pz(z)to a complex distribution Pg(x). Random noise is denoted by z. The target data sample is denoted by x. Generated sample distribution Pg(x) becomes indistinguishable from the real Pd (x) data by training the generator network. The discriminator is trained to identify the generated (fake) samples against data (real) samples using the minimax algorithm [10].

The GAN architecture used to generate the samples is Tacotron 2.

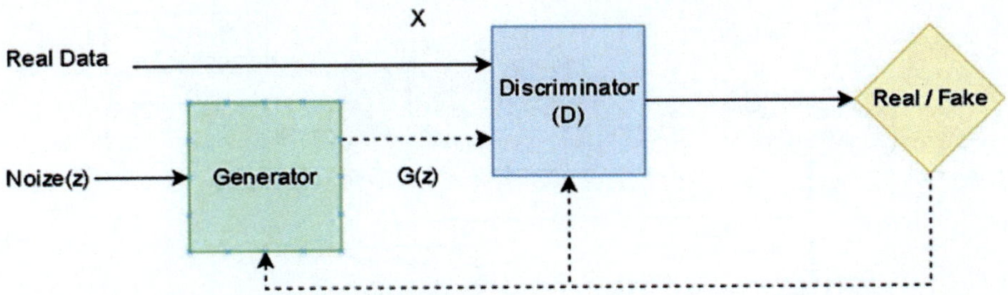

Fig. 1 Basic GAN architecture

Fig. 2 Tacotron 2 evolution

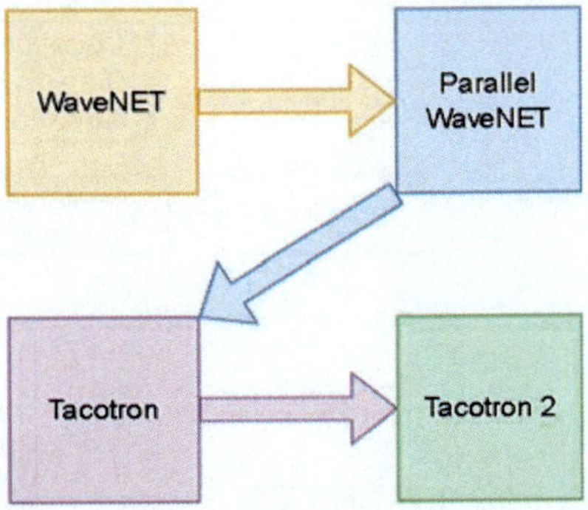

2.4 Tacotron 2

The image in Fig. 2 shows the evolution from Google WaveNET to Tacotron 2. [5, 11].

The Tacotron 2 architecture can be seen in Fig. 3. The Tacotron 2 architecture is an long short term memory (LSTM)-based Encoder-Attention-Decoder model. Tacotron 2 converts text to Mel spectrograms. Characters and phenomes are embedded using an encoder network. A convolutional stack processes the embedding. Subsequently, the output is sent to a bidirectional LSTM. An autoregressive LSTM acts as the decoder. On each call, a one-time slice of the Mel spectrogram is generated. The attention model connects the encoder and decoder. The attention model instructs the decoder which part of the encoded text to use. Tacotron 2 is essentially an recurrent neural network (RNN) and attention-based model that takes input text and produces a spectrogram. The spectrogram can be converted to speech by using a vocoder. Examples of vocoders are WaveNET or a classical Griffin-Lim algorithm. See Figs. 3 and 4.

In terms of model variations, various optimizers were tested, and it was found that the Adam optimizer provided the best results with fast model convergence. For every mini-batch, the gradient was set to zero before backpropagation avoiding the unnecessary accumulation of the gradient value. WaveNET and Griffin-Lim are tested for speech synthesis.

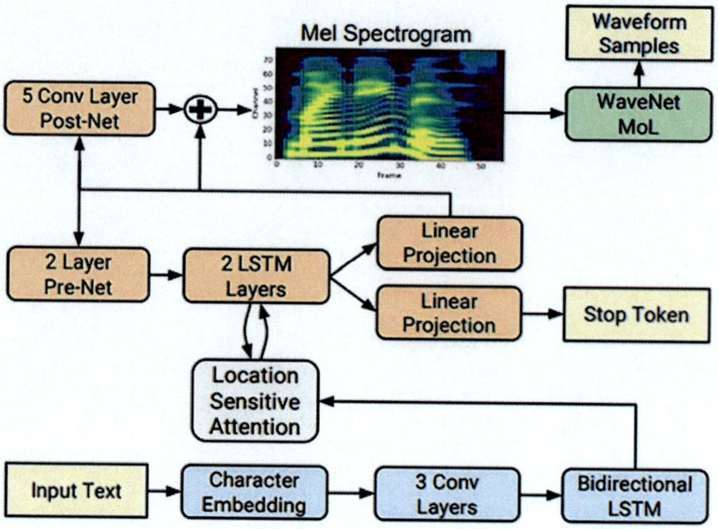

Fig. 3 Tacotron 2 System architecture using WaveNET vocoder [6]

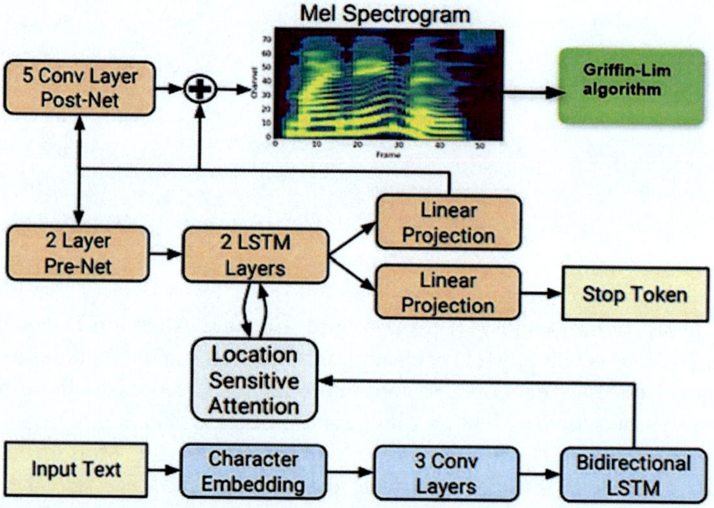

Fig. 4 Tacotron 2 System architecture using Griffin-Lim

2.5 *Database and Settings*

The dataset used to train the Tacotron 2 model is the NCHLT Afrikaans Speech Corpus. It contains an orthographically transcribed broadband speech corpus of approximately 56 h, including a test suite of 8 speakers [7]. The website can be seen in Fig. 5.

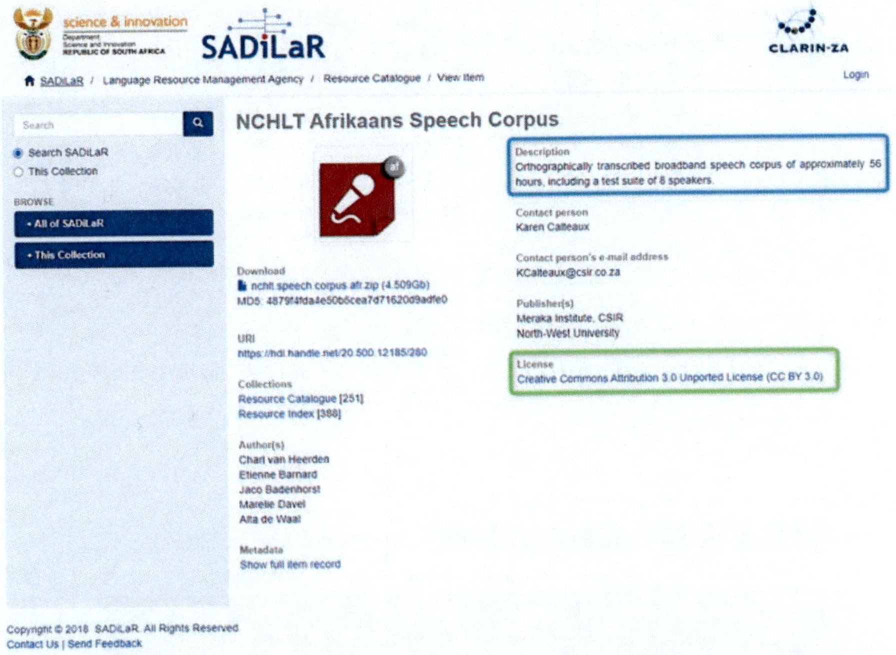

Fig. 5 Screenshot of website [7]

Fig. 6 Training files and sizes

The screenshot of the files and sizes are shown in Fig. 6.

3 Evaluation

3.1 Settings

From a hardware perspective, the Google Colab and the Kaggle platform are used for training. Access to high-end GPUs like the P100, V100, and T4 is available. The deep learning frameworks utilized are Tensorflow and PyTorch. In total, there are 66,133 files in the training data as can be seen in Fig. 7. Files are sampled at 22050 hz.

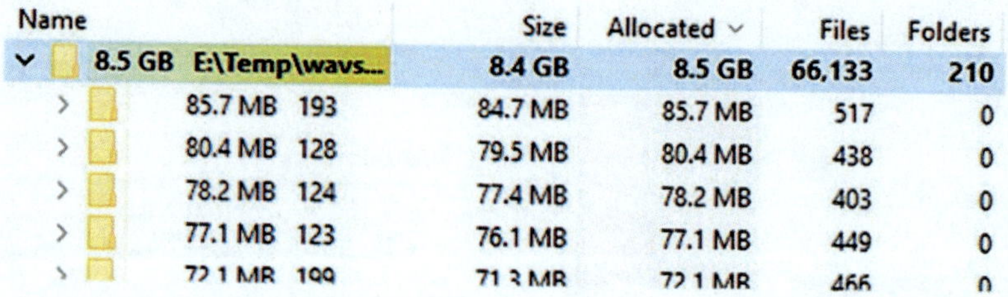

Name		Size	Allocated	Files	Folders
8.5 GB	E:\Temp\wavs...	8.4 GB	8.5 GB	66,133	210
85.7 MB	193	84.7 MB	85.7 MB	517	0
80.4 MB	128	79.5 MB	80.4 MB	438	0
78.2 MB	124	77.4 MB	78.2 MB	403	0
77.1 MB	123	76.1 MB	77.1 MB	449	0

Fig. 7 Training data

Fig. 8 Epoch training

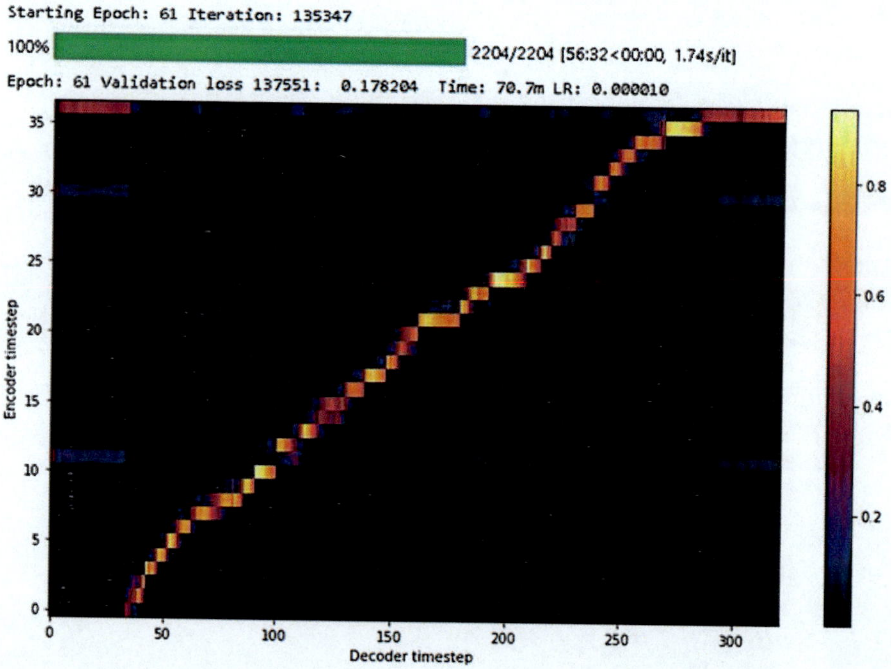

Fig. 9 Tacotron 2 Network Training

3.2 Training

Training time per epoch is roughly 1 h as seen in Fig. 8.

The .wav files are converted to. npy Numpy arrays containing the Mel spectrograms and then the Tacotron 2 network is trained. Once validation loss approaches 0.15, the model is ready to use. The training process can be seen in Fig. 9.

```
speech_to_speak_1 = "Julle maak my kwaad" #@param {type:"string"}
for index in range(5):
    end_to_end_infer(speech_to_speak_1, pronounciation_dictionary, show_graphs)
```

```
Current Config:
pronounciation_dictionary: False
show_graphs: False
max_duration (in seconds): 0.1
stop_threshold: 0.1

Warning! Reached max decoder steps
```

▶ 0:04 / 0:05

Fig. 10 Generating speech using trained Tacotron 2 model

Fig. 11 Plutchnik emotions

- Anger
- Anticipation
- Disgust
- Fear
- Joy
- Sadness
- Suprise
- Trust

The code used NVIDIA Tacotron 2 model [11, 13]. The code has been adapted and changed to cater to the Afrikaans speech corpus.

3.3 *Synthesizing Speech from Text*

The trained model is now used to generate speech clips. This can be seen in Fig. 10.

Audio clips are generated using various phrases [14]. These are categorized into the various emotion categories of Anger, Anticipation, Disgust, Fear, Joy, Sadness, Surprise, and Trust. See Figs. 11 and 12.

Using the WaveNET vocoder provides superior-quality voice clips when compared with the Griffin-Lim vocoder.

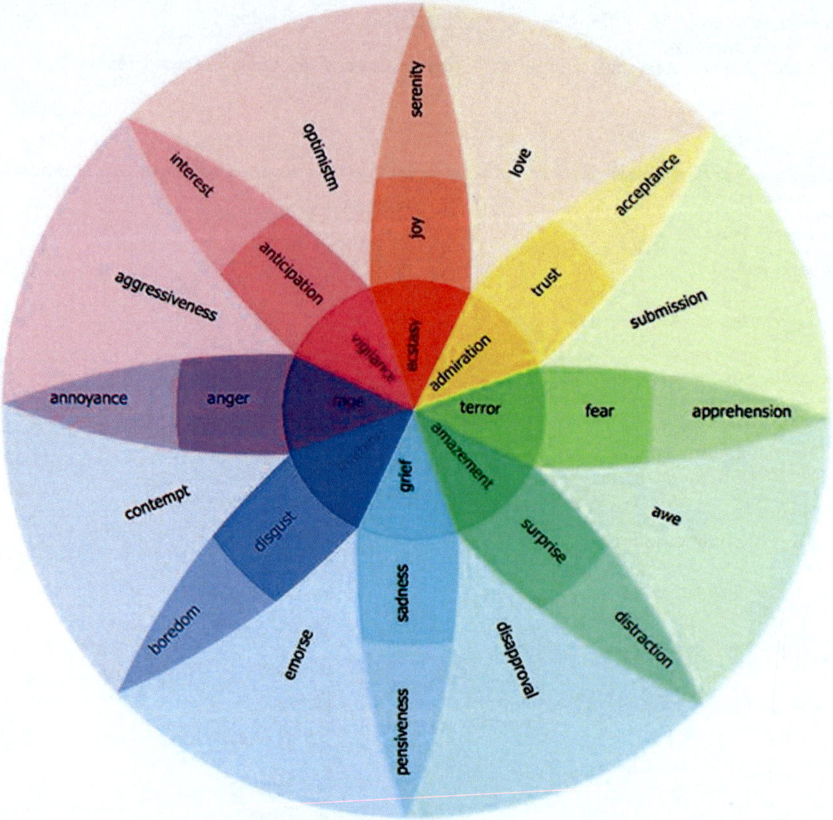

Fig. 12 Plutchnik wheel of emotions [15]

When comparing speech generated by the new model to other cloud providers like Google Cloud Text-to-Speech and Narakeet quality is the same. One gets a natural-sounding voice. Commercial Afrikaans text-to-speech providers are limited in comparison to other languages.

A mean opinion scale (MOS) has been the recommended measure of text-to-speech quality (ITU-T P.85, 1994) consisting of seven 5-point scales that assess overall sound quality, listening effort, comprehension problems, articulation, pronunciation, speaking rate, and pleasantness [15]. The 5-point scale has the following ratings: 1 – Bad; 2 – Poor; 3 – Fair; 4 – Good; and 5 – Excellent. The scoring of the Afrikaans tacotron 2 model compared to Google Cloud Text-to-Speech can be seen in Table 1 (Mean Opinion Score (MOS))..

Table 1 MOS

Model	Afrikaans Tacotron Griffin-Lim	Afrikaans Tacotrons WaveNET	Google Cloud Text-to-speech
Global impression	2	4	5
Listening effort	3	4	5
Comprehension problems	2	5	5
Speech sound articulation	3	5	5
Pronunciation	2	5	5
Speaking rate	2	4	5
Voice pleasantness	1	4	5

4 Conclusion

The current study has proposed generating synthetic speech samples specifically for the Afrikaans language. Various frameworks were investigated and it was found that the Tacotron 2 is the most advanced and fastest model to use. Vocoder variations like Griffin-Lim and WAVNet were evaluated. The generated sound clips address the speech synthesis issues of naturalness, prosody, spontaneity, and ambiguities. In terms of a form of measurement, the MOS scale was used to do a quality comparison. The vocoder that yielded the best result is WaveNET. It must be said that the MOS for the Afrikaans model is slightly lower than the commercial offerings. Not a lot of commercial offerings for the Afrikaans language currently exist. The model will be made freely available to any researcher wishing to use and improve upon it.

Acknowledgements Thanks to Nvidia for supplying the open-source GitHub source code that allowed the model to be trained and the speech samples to be created. This research is partially supported by the South African National Research Foundation (Grant Nos. 132797 and 137951), the South African National Research Foundation incentive grant (No. 114911), and the South African Eskom Tertiary Education Support Programme.

References

1. R. Yamamoto, E. Song and J. M. Kim, "Parallel Wavegan: A Fast Waveform Generation Model Based on Generative Adversarial Networks with Multi-Resolution Spectrogram," ICASSP 2020–2020 IEEE International Conference on Acoustics, Speech and Signal Processing, pp. 6199–6203, 2020.
2. Q. Tian, X. Wan and S. Liu, "Generative Adversarial Network based Speaker Adaptation for High Fidelity WaveNet Vocoder," Computer and information sciences, 2018.
3. B. H. Story, "History of Speech Synthesis," The Routledge Handbook of Phonetics, p. 9–33, 2019.
4. J. Shen and R. Pang, "Tacotron 2: Generating Human-like Speech from Text," 19 12 2017. [Online]. Available: https://ai.googleblog.com/2017/12/tacotron-2-generating-human-like-speech.html.

5. J. Shen, R. Pang, R. J. Weiss, M. Schuster, N. Jaitly, Z. Yang, Z. Chen, Y. W. Yu Zhang, R. Skerry-Ryan, R. A. Saurous, Y. Agiomyrgiannakis and Y. Wu, "Natural TTS Synthesis by Conditioning WaveNet on Mel Spectrogram Predictions," ICASSP 2018, 2017.
6. P. Salza, E. Foti, L. Nebbia and M. Oreglia, "MOS and Pair Comparison Combined Methods for Quality Evaluation of Text-to-Speech Systems," Acta Acustica united with Acustica, vol. 82, pp. 650–656, 07 1996.
7. R. Nielek, M. Ciastek and W. Kopeć, "Emotions Make Cities Live," Proceedings of the International Conference on Web Intelligence, 2017.
8. N. NGC, "NVIDIA NGC Catalog," [Online]. Available: https://catalog.ngc.nvidia.com/orgs/nvidia/teams/tlt-jarvis/models/speechsynthesis_english_tacotron2. [Accessed 19 01 2023].
9. F. Ma, Y. Li, S. Ni, S.-L. Huang and L. Zhang, "Data Augmentation for Audio-Visual Emotion Recognition with an Efficient Multimodal Conditional GAN," Applied Sciences, vol. 12, no. 1, p. 527, 2022.
10. J. Liu, C. Zhang, Z. Xie and G. Shi, "A novel method for Mandarin speech synthesis by inserting prosodic structure prediction into Tacotron2," International Journal of Machine Learning and Cybernetics, vol. 12, no. 10, p. 2809–23, 2021.
11. Y. Kumar, A. Koul and C. Singh, "A Deep Learning Approaches in Text-To-Speech System: A Systematic Review and Recent Research Perspective," Multimedia Tools and Applications, pp. 1573–7721, 12 September 2022.
12. K. Kuligowska, P. Kisielewicz and A. Włodarz, "Speech Synthesis Systems: Disadvantages and Limitations," International Journal of Engineering & Technology, vol. 7, no. 2, p. 234, 2018.
13. A. A. Karim and S. M. Saleh, "Text to speech using Mel-Spectrogram with deep learning algorithms," Periodicals of Engineering and Natural Sciences, vol. 10, no. 3, pp. 380–386, June 2022.
14. C. v. Heerden, E. Barnard, J. Badenhorst, M. Davel and A. d. Waal, "NCHLT Afrikaans Speech Corpus." Audio Recordings Smartphone-Collected in Non-Studio Environment," 2014. [Online]. Available: https://repo.sadilar.org/handle/20.500.12185/280. [Accessed 19 1 2023].
15. I. Goodfellow, J. &. M. Pouget-Abadie, B. Mehdi & Xu, D. &. O. Warde-Farley, S. &. Courville and Y. Aaron & Bengio, "Generative Adversarial Networks," Computer and information sciences, 2014.
16. D. Ferris, "Techniques and Challenges in Speech Synthesis," 2017.

Prediction of Bee Population and Number of Beehives Required for Pollination of a 20-Acre Parcel Crop

Yukun Jin, Tianyi Wei, Jingru Shi, Tingwen Chen, and Kai Yang

1 Introduction

The decline of the bee population has already become common sense [1–3]. According to FAO (Food and Agriculture Organization of the United Nations), bees are the most popular pollinator and affect 35% of the world's crop production, increasing outputs of 87% of the leading food crops worldwide, while there has been a drastic decrease in their population [4]. For instance, in Europe, 9.2% of bees are under threat of extinction [5]. Bees are certainly vital for the agricultural industry, since they, as the most popular pollinator, may contribute to the production of agricultural products with great economic returns, such as oil-bearing crops, nuts, and fruits [6–8]. In 2000, in the United States, the value of the increased yield and quality achieved through pollination by honey bees alone was about $14.6 billion [9]. Naturally, it is important to make predictions of bee populations using mathematical approaches [10]. Here, we'll use mathematical models to solve three problems related to the bee population.

Y. Jin · T. Wei · J. Shi · T. Chen
Amazingx Academy, Foshan, China

K. Yang (✉)
Sanya Science and Education Innovation Park, Wuhan University of Technology, Sanya, China

Xian Institute of Optics and Precision Mechanics of CAS, Xian, China

L. Meng (ed.), *International Conference on Cloud Computing and Computer Network*, Signals and Communication Technology,
https://doi.org/10.1007/978-3-031-47100-1_12

2 Bee Population Prediction Model A

2.1 Model Introduction

Our aim is to find the most reliable model to try to estimate the value of the number of bees based on the data we collected. We will try to find the error ratio of different prediction algorithms and find out the lowest one on average.

We decided to test the error ratio of four prediction algorithms using our database. The four algorithms contain three algorithms based on Grey Forecast Model and one of them is based on time series forecasting based on neural network modeling. For the Grey Forecast Model, we input seven sets of data for every model and form a model based on it. Then we compare the difference between the model and the data itself to find the error ratio. We've got three ways of Forecast models based on the Grey Forecast Model, and we test each of them individually using the same data to find out which one has the smallest error ratio which means a more accurate prediction. For the time series forecasting, we trained the neural network model using the first 10 datasets and tested it with the last one and the previous 9 datasets, then calculated the error ratio of its prediction. Comparing the error ratio of these four ways, we can find out which model best fits the data of the number of bees.

2.2 Data Collection and Processing

We collected the number of bees every year in every region from the official website of the food and agriculture organization [11]. The data contains the number of bees from 1961 to 2020. We choose 11 sets of data which arc showing a general trend as thc experimental data for both training models and identifying the error ratio [12, 13].

Because the data start at different ranges and the difference is too large, we standardized it first [14, 15]. Then we normalize the data into the range from 0 to 1 so we can use it to train neuron networks. We then linearly regress the data, compare the residual case order plot to find out the outliers in the data and delete them. We also used the three sigma rules to find out the extreme data in our database. After deleting all the abnormal data, we use the average of data to fill up those missing data. We also separate the 11 sets of data into training data and testing data to train and test the neural network model.

2.3 Model (1a)

1. *Grey Forecast Model.*
   (a) *DGM(2,1) Model.*
   When applying the DGM(2, 1) Model, we will have to first set up a time series.

$$x^{(0)} = \left(x^{(0)}(1), x^{(0)}(2), \ldots, x^{(0)}(n)\right)$$

the series' 1-AGO sequence $x^{(1)}$ and 1-IAGO sequence $\alpha^{(1)}x^{(0)}$ will be:

$$x^{(1)} = \left(x^{(1)}(1), x^{(1)}(2), \ldots, x^{(1)}(n)\right)$$

and

$$\alpha^{(1)}x^{(0)} = \left(\alpha^{(1)}x^{(0)}(2), \ldots, \alpha^{(1)}x^{(0)}(n)\right)$$

The model of DGM(2, 1) is

$$\alpha^{(1)}x^{(0)}(k) + ax^{(0)}(k) = b$$

Next construct a matrix B and vector Y:

$$B = \begin{bmatrix} -x^{(0)}(2) & 1 \\ -x^{(0)}(3) & 1 \\ \vdots & \vdots \\ -x^{(0)}(n) & 1 \end{bmatrix}, Y = \begin{bmatrix} \alpha^{(1)}x^{(0)}(2) \\ \alpha^{(1)}x^{(0)}(3) \\ \vdots \\ \alpha^{(1)}x^{(0)}(n) \end{bmatrix} = \begin{bmatrix} x^{(0)}(2) - x^{(0)}(1) \\ x^{(0)}(3) - x^{(0)}(2) \\ \vdots \\ x^{(0)}(n) - x^{(0)}(n-1) \end{bmatrix}$$

   (b) *GM(1,1) Model.*
   Let original series x0 be

$$X^{(0)} = \left\{X^{(0)}(i), i = 1, 2, \ldots, n\right\}$$

   Grey Model(1,1) performs well in predictions of sequences that increase exponentially.

   (c) *GM(2,1) Model.*
   For GM(2,1) Model, we first need to set up the initial array x0.In this situation, x0 is the array containing the number of bees in a chosen region every year.

Then we set up the first summation produced sequence:

$$(1-AGO)x^{(1)} = x^{(1)} = \left(x^{(1)}(1), x^{(1)}(2), \ldots, x^{(1)}(n)\right)$$

The function of data can be calculated:

$$\frac{d^2x^{(1)}}{dt^2} + a_1\frac{dx^{(1)}}{dt} + a_2x^{(1)} = b$$

$$B = \begin{bmatrix} -x^{(0)}(2) & -z^{(1)}(2) & 1 \\ -x^{(0)}(3) & -z^{(1)}(3) & 1 \\ \vdots & \vdots & \vdots \\ -x^{(0)}(n) & -z^{(1)}(n) & 1 \end{bmatrix}, Y = \begin{bmatrix} \alpha^{(1)}x^{(0)}(2) \\ \alpha^{(1)}x^{(0)}(3) \\ \vdots \\ \alpha^{(1)}x^{(0)}(n) \end{bmatrix} = \begin{bmatrix} x^{(0)}(2) - x^{(0)}(1) \\ x^{(0)}(3) - x^{(0)}(2) \\ \vdots \\ x^{(0)}(n) - x^{(0)}(n-1). \end{bmatrix}$$

So we can calculate the:

$$\widehat{u} = \begin{bmatrix} \widehat{a_1} \\ \widehat{a_2} \\ b \end{bmatrix}$$

and plug-in a1 a2, and b to get the function.

2. *Time Series Forecasting Model.*

In this problem, because our data is using time as an independent variable, we can use the time series, forecasting model. Data received at different times that describe the change of one or more features over time is time series data, and the time series forecasting model tries to predict future data based on existing time series data. Since the change in the number of bees is not directly related to the years, we can use the NAR (Nonlinear Auto Regressive Neural Network) to predict the data. To alleviate the problem of too strong an independence assumption, a scheme is to introduce the hidden variable z and get:

$$P_\theta(y|x) = \int_z P_\theta(y|z,x)p_\theta(z|x)dz$$

Assuming that the target sequence is independent under the premise of given hidden variables, then:

$$P_\theta(y|z,x) = \prod_{t=1}^{T} P_\theta(y_t|z,x)$$

This is the basic idea of NAR neural network, and it can use this algorithm to predict the future terms in the target sequence.

We trained our model using 10 sets of data, including 70% for training, 15% for verification, and 15% for testing. We test our model out when the training is finished.

2.4 Evaluation (1a)

1. *Grey Forecast Model.*
   To find the error ratio of the model, we can compare the result of the prediction and the real value. Using this program, we tested seven sets of data for three different methods including DGM(2,1), GM(1,1), and GM(2,1). After calculating the error ratio of each of them, we got Tables 1 and 2 and diagram below.
   As a result, since the DGM(2,1) model has the lowest error ratio, we decided to test it with data from Low-Income Food Deficit Countries.
   The final Posterior error ratio is 0.2461 which is less than 0.35, showing that it can predict the bee numbers pretty accurately.

Table 1 Error ratio of the model in different regions by three methods

Number	Region	Posterior error ratio		
		DGM(2,1)	***GM(1,1)***	***GM(2,1)***
1	World	0.23328	0.23707	0.2633
2	Western Asia	0.11207	0.26088	0.11673
3	Cameroon	0.11985	0.39319	0.09286
4	Turkiye	0.13627	0.24392	0.15839
5	Tunisin	0.32124	0.98804	1.2879
6	Asia	0.14113	0.15042	0.22573
7	Least developed countries	0.13303	0.21512	0.14573
8	Average	0.170981	0.35552	0.32723

Table 2 Original data and predicted data through time

Year series number	Original	Prediction
1	0.7528	0.6033
2	0.788	0.631
3	0.7685	0.6597
4	0.8527	0.6896
5	0.8586	0.7207
6	0.8921	0.7529
7	0.9372	0.7865
8	0.9632	0.8213
9	0.9891	0.8576
10	0.9894	0.8952

2. *Time Series Forecasting Model.*
   We found that most of the error ratio is from -0.1 to 0.1, showing that it can predict the value mostly precisely.
3. *Conclusion*
   By comparing the error ratio, we can conclude that time series prediction is more accurate than grey forecast modeling, but it needs a large database to set up the model. When we are trying to predict the population of bees in a small region with less data, we can still use Grey Forecast modeling, especially using the DGM(2,1) model since it has significantly higher accuracy than other methods.

3 Bee Population Prediction Model B

3.1 Model Introduction

Using a mathematical model to predict the population of bees, the model should express populations in terms of several parameters and give predictions as accurately as possible. A differential model will be applied here. Here, 10 variables would be considered during the prediction. Next, we will simplify the equation, so that it can express the future population. At last, we will plug the value of time into the equation to obtain the predicted population.

3.2 Data Collection and Processing

For this method, we'll only need one set of data, so we'll simply pick a set of data from data sets collected for model A for problem 1.

3.3 Model (1b)

First, based on the growth rate and the decline rate, we can get the increase rate of population based on the survival number of eggs and the declining number based on natural and environmental factors.

$$\frac{dP(t)}{dt} = fe * Rs(t) - P(t) * fd - a * P(t)^2$$

The first term is the number of survival eggs increased. The second term is the number of the population declined by natural factors that are directly proportional to the population itself. The third term is the environmental limitation that when the population is big enough, the increase rate will greatly decline and finally become zero. a is a constant that can be calculated by the information given.

Then we can get the function of the survival rate at time t which is:

$$Rs(t) = k * dT * Vf * P(t)$$

Since, in the assumption, we simply give a proportional relationship between the survival rate and those three variables containing the change rate of temperature, the relative speed of flying, and the current population, we can directly use a constant k to set the relationship.

$$\frac{dP(0)}{dt} = R \frac{dP(T_{\max})}{dt} = 0$$

We then set up the start and end situation so that at time 0, it has a survival rate of R which should be pretty high since the environmental factor is not affecting that much, and at time Tmax when the population is at its highest point, and should not go up anymore.

3.4 Evaluation (1b)

The Posterior difference ratio is 0.1865 which means that it can fit the value even better than DGM(2,1). Both of the ways to forecast the bee population are precise enough and we can use different methods based on the information we got. If we have more information about the bee itself, it's better to set up a model using method 1b, since it won't need a lot of other data other than the bee and the environment itself.

3.5 Results for Both Models

Both model a(DGM(2,1)) and model b give results showing that the population of bees will increase in the next 10 years. However, the predicted rates of increase are different. Model a gives a lower increase rate, and it is steady, while Model b gives a higher rate of increase with increasing gradient. The total number of bees increased in Food Deficit countries in 10 years will be approximately 3,234,000 and 5,428,000, according to Model a and Model b, respectively. (See "The graph of bee number over time passed")

4 Bee Population Sensitivity Test

4.1 Model Introduction

For this question, we aim to find a reliable model to discover the relationship between variables like characteristics of bees or climate and the colony size of bees by calculating the ratio between percentage change in bee population resulting from a percentage change in variables.

Then, we will use the Random forest model to find the importance of variables we didn't contain in the function.

4.2 Data Collection and Processing

We first categorize the bees into four types: Apis Cerana Fabricius in India, Apis Mellifera Linnaeus in Chad, Apis Cerana Fabricius in Turkey, and Apis Mellifera Linnaeus in Cameroon. Next, we gather 10 sets of variables that may affect the population of bees for each type. Most of them are gathered from FAO, and the rest of them are from World Data Bank and Wikipedia.

4.3 Model (2a)

1. *Algorithms of random forest.*
   (a) *Decision trees have to be set up for preliminary in two possible ways.*
      - By boosting (serial algorithm such as AdaBoost algorithm).
      - By bagging (bootstrap aggregating).

      The training algorithm for random forests applies the general technique of bootstrap aggregating, or bagging, to tree learners. Given a training set X = x1, ..., xn with responses Y = y1, ..., yn, bagging repeatedly (B times) selects a random sample with replacement of the training set and fits trees to these samples:

$$\hat{f} = \frac{1}{B}\sum_{b=1}^{B} f_b(x')$$

      or by taking the majority vote in the case of classification trees.

      So that our algorithm performs better because of the decreased variance of the model.

(b) *Random forest.*

Classical decision trees choose the best property from the property set (assume there are d properties); however, for the Random Forest, for every junction of the tree, a subset that contains k properties is selected randomly from the property set at that junction, then the best property is selected from that subset.

We introduce parameter k to control the degree of randomness. if $k = d$, the construction of the decision tree is the same as the classical one. For most circumstances, $k = \log2(d)$ is used.

For a set that contains T base learner{h,h2,...,hT}, in which hi(x) is the output, two methods can be used:

- Voting(majority voting, plurality voting, and weighted voting).
- Averaging.

2. Differential *function* analysis.

In this method, we will set all our variables to their estimated value as an initial value, and change each variable by 100% to see the change in the result population.

4.4 Evaluation

1. *Random forest result.*

By testing four sets of data including India, Chad, Turkey, and Cameroon, we analyzed eight variables that might influence the population of bees.

The final model has an R square of 0.89, showing that it can predict and fit our data really precisely, so the final result is reliable.

We found that the year, which is the time passed, is definitely the most important variable. Apart from time, the initial population also affects the honey bee population largely, since it can affect the rate of change in the population a lot.

4.5 Function Analysis Result

We test the sensitivity of each variable by changing the variable by 100% and comparing the output of the function with the original value. We choose the original time as 5 years.

The time and original population influence the result the most, as shown in Fig. 1, other variables such as flying speed or change in temperature didn't matter that much. The original rate of living and the constant death rate is also important. But the original rate of living is less important as time passed, and the effect of this rate decreased a lot.

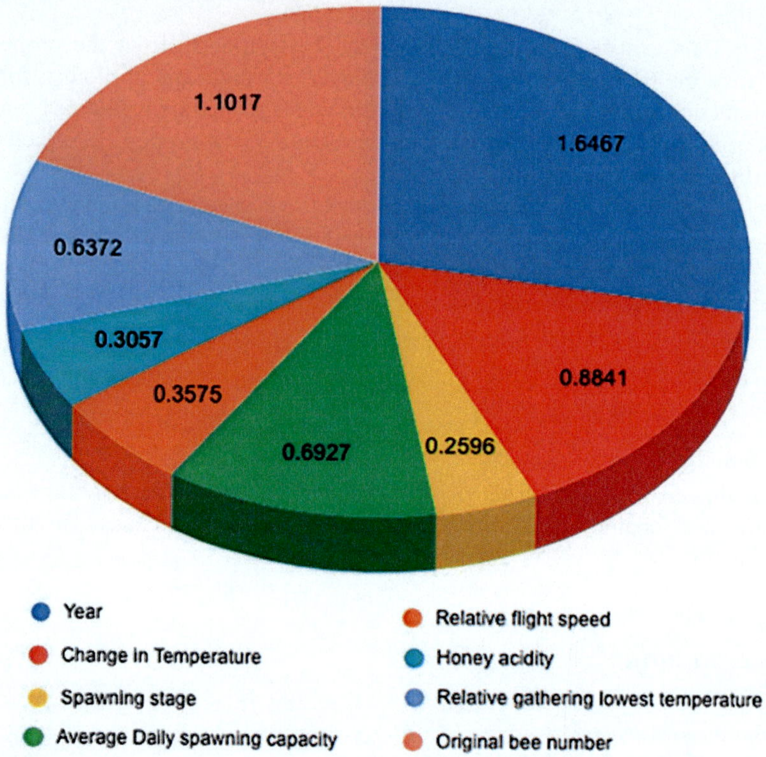

Fig. 1 Variable importance

5 Beehive Number Estimation Model

5.1 Model Introduction

To find the least number of bee hives required to support the pollination of a 20-acre parcel of land, the models should be able to consider two different types of arrangement of bee hives and give the arrangement with the least number of bee hives needed. We express the possibilities for a flower to be pollinated at a certain distance by considering the maximum number of flowers a bee can serve and the density of flowers, and assuming that the possibility of a flower getting pollinated has a normal distribution relationship with the distance it keeps from the hive.

5.2 Data Collection

Here, we will have to consider the influence of time on the population of bees. This variation of the population with time could be predicted by model 1b. The other data needed is given by the stem of problem B3. (The area of pollination)

5.3 Model

By using model 1b, we can get the function of the population through time. We can use the function to find out the time it takes to get to its highest population and the number of its highest population. Then we can change the σ of the normal distribution function. The σ should be limited to a specific range. By observing the graph, 0.2–1.2 should be the best range if the maximum distance a bee will travel is 6 km.

5.4 Evaluation

In this specific problem, we find out the population function that is:

$$P(t) = \frac{20}{0.04 * e^{-7*t} + 0.000335} + 20000$$

By using this function and the value provided in the problem, we can find the function of possibilities at a point x meters from the beehive. Here we use the crop cron as an example, the specific value of its planting density is 4.5 per meter square.

$$a = \frac{P(t) - 20000}{60000} + 0.2$$
$$k = \frac{P(t) * 2000}{\pi * 6000^2 * 4.5}$$
$$p(x) = \frac{e^{-\frac{\frac{x^2}{3}}{2 \cdot a^2}}}{\sqrt{2\pi} \cdot a} * k$$

By setting up a matrix to find the value of possibility at each point, we get the result below. We can see that two beehives are already enough for a field 20-acre if we want a possibility just bigger than 80%.

5.5 Results

Two beehives are already enough for a field 20-acre if we want a possibility of pollination for plants just bigger than 80%.

Acknowledgments This work was in part supported by Project of Sanya Yazhou Bay Science and Technology City (Grant No: SCKJ-JYRC-2022-17) and Sanya Science and Education Innovation Park of Wuhan University of Technology (Grant No:2022KF0020).

References

1. A., N. (2014). European Red List of Bees. Publication Office of the European Union. https://policycommons.net/artifacts/1374072/european-red-list-of-bees/1988308/
2. Food and Agriculture Organization of the United Nations. (2019, May 20). News Article: Declining bee populations pose threat to global food security and nutrition. FAO. Retrieved November 5, 2022, from https://www.fao.org/news/story/en/item/1194910/icode/
3. Morse, R. A., & Calderone, N. W. (2000). The value of honey bees as pollinators of US crops in 2000. Bee culture, 128(3), 1–15.
4. Collins, A. M., Rinderer, T. E., Harbo, J. R., & Bolten, A. B. (1982). Colony defense by Africanized and European honey bees. *Science*, *218*(4567), 72–74.
5. Cairns, C. E., Villanueva-Gutiérrez, R., Koptur, S., & Bray, D. B. (2005). Bee Populations, Forest Disturbance, and Africanization in Mexico 1. *Biotropica: The Journal of Biology and Conservation*, *37*(4), 686–692.
6. Meixner, M. D. (2010). A historical review of managed honey bee populations in Europe and the United States and the factors that may affect them. *Journal of invertebrate pathology*, *103*, S80–S95.
7. Hatjina, F., Costa, C., Büchler, R., Uzunov, A., Drazic, M., Filipi, J., ... & Kezic, N. (2014). Population dynamics of European honey bee genotypes under different environmental conditions. *Journal of Apicultural Research*, *53*(2), 233–247.
8. Gezon, Z. J., Wyman, E. S., Ascher, J. S., Inouye, D. W., & Irwin, R. E. (2015). The effect of repeated, lethal sampling on wild bee abundance and diversity. *Methods in Ecology and Evolution*, *6*(9), 1044–1054.
9. Mallinger, R. E., Gaines-Day, H. R., & Gratton, C. (2017). Do managed bees have negative effects on wild bees?: A systematic review of the literature. *PloS one*, *12*(12), e0189268.
10. Mallinger, R. E., Gaines-Day, H. R., & Gratton, C. (2017). Do managed bees have negative effects on wild bees?: A systematic review of the literature. *PloS one*, *12*(12), e0189268.
11. Bryden, J., Gill, R. J., Mitton, R. A., Raine, N. E., & Jansen, V. A. (2013). Chronic sublethal stress causes bee colony failure. *Ecology letters*, *16*(12), 1463–1469.
12. Rollin, O., Bretagnolle, V., Decourtye, A., Aptel, J., Michel, N., Vaissière, B. E., & Henry, M. (2013). Differences of floral resource use between honey bees and wild bees in an intensive farming system. *Agriculture, Ecosystems & Environment*, *179*, 78–86.
13. Genersch, E. (2010). Honey bee pathology: current threats to honey bees and beekeeping. *Applied microbiology and biotechnology*, *87*, 87–97.
14. Chai-Ead, N., Aungkulanon, P., & Luangpaiboon, P. (2011, March). Bees and firefly algorithms for noisy non-linear optimisation problems. In *Proceedings of the international multi conference of engineering and computer scientists* (Vol. 2).
15. Berry, J. A., Hood, W. M., Pietravalle, S., & Delaplane, K. S. (2013). Field-level sublethal effects of approved bee hive chemicals on honey bees (Apis mellifera L). *PloS one*, *8*(10), e76536.

Index

L. Meng (ed.), *International Conference on Cloud Computing and Computer Network*, Signals and Communication Technology,
https://doi.org/10.1007/978-3-031-47100-1